Cooperative Effects in Stochastic Models

COOPERATIVE EFFECTS IN STOCHASTIC MODELS

G. SH. TSITSIASHVILI AND M.A. OSIPOVA

Nova Science Publishers, Inc.

New York

Library of Congress Cataloging-in-Publication Data

Tsitsiashvili, G. Sh. (Gurami Shalvovich)
 Cooperative effects in stochastic models / G. Sh. Tsitsiashvili and M.A. Osipova.
     p. cm.
Includes index.
ISBN 1-59454-252-X
1. Economics--Mathematical models. 2. Stochastic analysis. I. Osipova, M. A. II. Title.
HB135.T77 2004
330'.01'51922--dc22
                            2004030808

Copyright © 2005 by Nova Science Publishers, Inc.
                     400 Oser Ave, Suite 1600
                     Hauppauge, New York 11788-3619
                     Tele. 631-231-7269   Fax   631-231-8175
                     e-mail: Novascience@earthlink.net
                     Web Site: http://www.novapublishers.com

Printed in the United States of America

Contents

Introduction   **vii**

1   Cooperative effects in discrete queueing models   **1**

    1.1.   Objective functions of $M/M/1/\infty$ unities   1

    1.2.   Product form stationary distributions of Markov processes . .   11

    1.3.   Product theorems for queueing networks
with switching centers   17

    1.4.   Product theorems for queuing networks with prohibitions . . .   22

    1.5.   Stochastic control of Markov process parameter   30

2   Cooperative effects in risk models with discrete time   **35**

    2.1.   Model of mutual insurance with independent risks   36

    2.2.   Mutual insurance models with weak dependent risks   58

3   New approaches to heavy tails asymptotic analysis   **67**

    3.1.   Investigation of applied probability models by means of random variables indexes   68

    3.2.   Multiserver queueing system with competition between servers   76

4   Cooperative effects in reliability models   **87**

    4.1.   Renewal systems with common reserve   88

    4.2.   Effectiveness of separate reserve   97

5   New approaches to data processing problems   **101**

    5.1.   Growth models   101

    5.2.   Optimal algorithm of discrete images reconstruction   109

5.3.   Layering method in biorhythmics problems 110

5.4.   Stationary random processes with finite correlation interval . 112

5.5.   Estimates of parameters in parallel and sequential connections of independent elements 116

5.6.   Estimate of exponential distribution parameter 117

5.7.   Recognition of extremal objects by interval decision rule . . . 118

6 Cooperative effects in stochastic models of mechanics   121

6.1.   Anomalous Diffusion with Periodical Initial Conditions on Interval with Reflecting Edges 121

6.2.   Random time of threads wisp break 135

6.3.   Construction and investigation of stochastic model of sliding regime 139

7 Cooperative and nonlinear effects in models of mathematical economics   147

7.1.   Cooperative effects in the simplest stochastic growth model . 148

7.2.   Basis and development of algorithm of mutual debts discharge solution 150

7.3.   Qualitative investigation of Estes self-education model 164

Bibliography   171

Index   179

Introduction

This monograph is devoted to an investigation of cooperative effects in stochastic models. It includes original results of authors over the last decade. Main object of the monograph is an analysis of the influence of stochastic model structure on its characteristics.

Problems of cooperation and a decomposition are actual in the solution of many concrete problems. These problems are: parallelization of algorithms and programs, modelling of supercomputers, computer networks, systems of mobile telephones, catastrophes in complex systems, a design and an improvement of technological and economical processes and etc.

The cooperative effects create a source of significant dependencies between complex system characteristics under large random disturbances. To analyze these effects, it is necessary to create special methods based on structural analysis of multi-element stochastic models together with majoral asymptotic bounds of these model's characteristics. At the same time, it demands developing new approaches to the processing of statistical data and skill in usage of the probability theory limit theorems and related asymptotic series and bounds.

A choice of the monograph material is defined as by initial applied problems and by probability methods of their solution. Conditionally the monograph may be divided into two parts. The first of them contains four sections devoted to finding cooperative effects and to the development of new related analytical and numerical methods. This part has a presumably methodological character and creates a theoretical base of an investigation of applied stochastic systems. The second part contains three sections devoted to a solution of different applied problems. It has some interesting substantial results.

In the first section cooperative effects in discrete queuing models are considered. Different regimes of a counteraction between elements of complex queuing systems are analyzed. The choice of models is connected with a sufficient simplicity of calculations, asymptotic bounds and proofs of the main statements. The section contains a few new product theorems for queuing networks theory.

The second section is devoted to an investigation of the cooperative effects in models of mutual insurance and their analogs in queuing theory. In the models of mutual insurance, the phenomenon of so-called "the phase transition" for limit characteristics is found. All calculations of this section are based on the central limit theorem and estimates of the large deviations in this theorem and on some other elements of probability theory.

The third section contains new approaches to asymptotic analysis of distributions with heavy tails. This analysis is a fast developing topic of probability theory. Methods suggested in this section allow separating some new cooperative effects in risk and queuing models and analyzing objective functions in multiserver queuing systems with competition between servers.

In the fourth section, the cooperative and decomposition effects in static and dynamic reliability models are considered. An analytical consideration is connected here with numerical calculations. These calculations help to separate the effects and to estimate accuracy of related asymptotic bounds.

The fifth section consists of a series of different statistical algorithms and bounds which have been obtained using cooperative effects considered in the previous sections. To design these algorithms, some nontraditional statistical approaches were created and used: special methods of data organization in concrete considerations and special statistical estimates of unknown parameters for random processes.

The sixth section is devoted to an extraction of the cooperative effects in some stochastic mechanical systems. Numerical investigation of these models was made presumably thanks to the introduction of large parameters characterized by a number of subsystems into these systems and their models.

In the seventh section, the cooperative and nonlinear effects in mathematical economics models are considered. Using a technique of linear programming, functional analysis and probability theory, some interesting questions of operations research in economics systems are considered.

The authors thank their teachers – RAS Academician E.V. Zolotov, RAS corresponding member V.P. Korobeinikov, professors, doctors and doctors of science, V.M Zolotarev and V.V. Kalashnikov, whose ideas are used in this monograph. Further, the authors thank professors and doctors of science, D. Konstantinides (Greece), D. Baum (Germany), Q. Tang (China), L.Ya Glybin, R.G. Barantsev, and D.S. Anikonov for cooperation in the solution of some concrete problems considered in the monograph. The authors also thank candidate of sciences V.A. Sviatukha, Yu. N. Kharchenko and V.M. Bespalov for their useful discussions.

The authors hope that this monograph increases interest in problems of stochastic modelling and help young researchers to choose a direction for their investigations.

Chapter 1

Cooperative effects in discrete queueing models

In this chapter, the effects of an aggregation of queueing systems, described by discrete Markov processes, are investigated. These models appear like "simple" ones but an investigation of the cooperative effects in them is new and serves as a source for new results. These results include as cooperative effects so their dependence on a choice of objective functions and on the way of the system aggregation.

A presence of strong counteraction or an absence of the counteraction are two opposite and extreme cases in the aggregation. But even these extreme regimes have a lot of different modifications which were not considered earlier.

In this chapter, special attention is devoted to a calculation of product stationary distributions in considered models. The possibility to transit from the product form to the conditional product form of the stationary distributions in stochastic models is analyzed. This new form corresponds to an intermediate case between the strong dependence and the complete independence of system elements in stationary regimes.

1.1.   Objective functions of $M/M/1/\infty$ unities

In this subsection queuing systems composed as unities of identical or nonidentical one-server queuing systems $M/M/1/\infty$ into multi-server or one-server queuing system are considered. The effects of a such aggregation are significant if a number of aggregated systems is large or if their load is high so that it is possible to say about a strong counteraction between the united systems. This circumstance allows to build from oneserver queuing systems different aggregations with the characteristics which are significantly better than the characteristics of the initial systems. A variant of such a construction is suggested here.

Identical servers

Aggregate the identical oneserver queuing systems $M/M/1/\infty$ into the multiserver queuing systems with a stationary queue length and a waiting time which are much smaller than analogous characteristics of their components. Investigate quantitatively how a way of the aggregation influences on these objective functions of the final queuing systems.

Denote $X_1, X_2, \ldots$ the sequence of the independent queuing systems $M/M/1/\infty$ with the input and the serving intensities λ and μ which satisfy the following ergodicity condition:

$$\rho = \frac{\lambda}{\mu} < 1. \tag{1.1}$$

Construct from n elements of this sequence a such queuing system S in which the stationary queue length $a(S)$ and the characteristic time (of an input into stationary regime) $\tau(S)$ satisfy the inequalities

$$a(S) \leq A, \quad \tau(S) \leq T.$$

Here A, T are some positive numbers which are significantly smaller than the analogous characteristics $a(X_i), \tau(X_i)$ of the initial sequence elements. Such a formulation of the multicriteria problem becomes possible only after a recognition of the cooperative effects in queuing systems [82, 83], which give rules of the unification.

Consider the aggregation of the n independent one-server queuing systems $X_1, \ldots, X_n$, into the multiserver queuing system S_n organized as $M|M|n|\infty$ and represented at the figure 1 (here the needles denote the input and output flows, the circles show the servers with the queues and the rectangle – multiserver queuing system with the common queue):

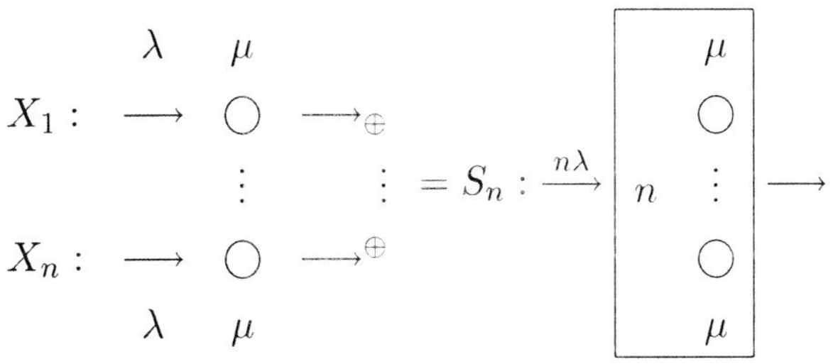

$$X_1: \quad \overset{\lambda}{\longrightarrow} \quad \bigcirc \quad \overset{\mu}{\longrightarrow}_\oplus$$

Fig. 1

This operation of the aggregation denotes by the formula

$$S_n = X_1 \oplus \ldots \oplus X_n.$$

As the ergodicity conditions (1.1) for the systems $X_1, \ldots, X_n$, S_n are true so for some positive C, α the mean stationary queue length $a(S_n)$ in the system S_n satisfies [82, c. 63] the inequality

$$a(S_n) \leq Ce^{-\alpha n}. \tag{1.2}$$

Consider the aggregation of the systems $X_1, \ldots, X_n$ into the sum $M|M|1|\infty$ showed at the figure 2:

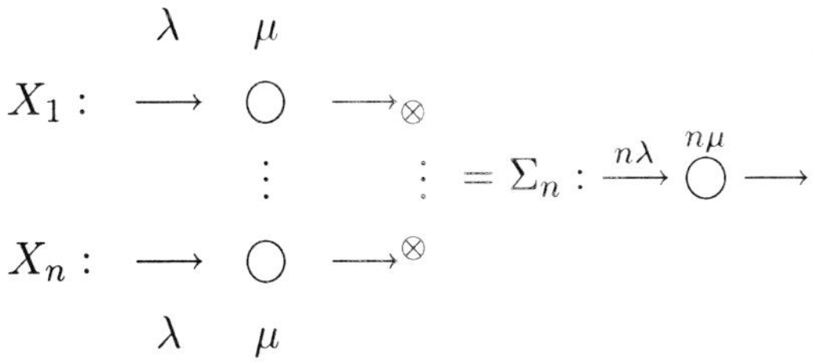

Fig. 2

The operation of the aggregation defined in such the way denotes by

$$\Sigma_n = X_1 \otimes \ldots \otimes X_n.$$

If the condition (1.1) is true then the mean stationary queue $a(\Sigma_n)$ in the system Σ_n satisfies the formula

$$a(\Sigma_n) \equiv a(X_1) < \infty. \tag{1.3}$$

Now learn how a form of the aggregation influences on the characteristic time. Suppose that $z(t)$ is the ergodic birth and death process with the intensities λ_i, μ_i, $i \geq 0$, $\mu_0 = 0$, and the distributions

$$p_i(t) = P\{z(t) = i\}, \quad p_0(0) = 1, \quad \pi_i = \lim_{t \to \infty} p_i(t), \quad i \geq 0.$$

Denote $d_{-1}, d_0, d_1, \ldots$ the sequence of the positive numbers so that

$$d_{-1} = d_0 = 1, \quad d_0 \sum_{i \geq 1} \pi_i + d_1 \sum_{i \geq 2} \pi_i + \ldots < \infty$$

and put

$$w_i(t) = \pi_i - p_i(t), \quad w_i(0) > 0, \quad v_i(t) = d_{i-1} \sum_{j \geq i} w_j(t),$$

$$a_i = \lambda_{i-1} + \mu_i - \frac{d_{i-2}}{d_{i-1}} \mu_{i-1} - \frac{d_i}{d_{i-1}} \lambda_i, \quad i \geq 1, \quad R = \sup_{i \geq 1} a_i, \quad r = \inf_{i \geq 1} a_i,$$

$$\vec{v}(t) = (v_1(t), v_2(t), \ldots), \quad v(t) = \sum_{i=1}^{\infty} v_i(t), \quad \|\vec{v}(t)\| = \sum_{i=1}^{\infty} |v_i(t)|.$$

In [97] (the formulas (8), (10), (11)) the following inequalities were obtained from the condition: there are $R, r > 0$, $0 < r < R < \infty$ so that

$$\|\vec{v}(t)\| \leq v(0)e^{-rt},$$

$$dv(t)/dt = -\sum_{i=1}^{\infty} a_i v_i(t) \geq -Rv(t).$$

If this condition is true then

$$v(0)e^{-Rt} \leq v(t) \leq \|\vec{v}(t)\| \leq v(0)e^{-rt}. \tag{1.4}$$

Apply the inequalities (1.4) to the system S_n in which the current number of the customers is described by the birth and death process with the intensities $\lambda_i = n\lambda$, $\mu_i = \mu \min(i, n)$, $i \geq 0$. Following [97] choose $d_i = \gamma^i$, $i \geq 0$, where $\gamma = 1 + c/n$, $0 < c < 1 - \rho$. It is clear that for $i \geq 0$ the following equalities are true

$$a_{i+1} = \lambda_i + \mu_{i+1} - \frac{\mu_i}{\gamma} - \gamma\lambda_{i+1} = \mu(-\rho c + b_i),$$

$$b_i = \min(i+1, n) - \frac{\min(i, n)}{\gamma} = \begin{cases} 1 + i(1 - 1/\gamma), \\ 0 \leq i < n, \\ n(1 - 1/\gamma), n \leq i. \end{cases} \tag{1.5}$$

Consequently

$$R = a_n \leq \mu(-\rho c + 1 + c(1 - 1/n)) \leq \mu(1 + c(1 - \rho)) = R_S < \infty, \qquad (1.6)$$

$$r = \min(a_1, a_{n+1}) = \mu(-\rho c + \min(1, n(1 - 1/\gamma))) \geq$$

$$\geq \mu c(1 - \rho - c) = r_S > 0. \qquad (1.7)$$

So the formulas (1.6), (1.7) guarantee for the system S_n the following inequality

$$v_S e^{-R_S t} \leq ||\vec{v}_S(t)|| \leq v_S e^{-r_S t}, \qquad (1.8)$$

$$v_S = \sum_{i \geq 1} \gamma^{i-1} \sum_{j \geq i} \pi_j < \infty, \quad ||\vec{v}_S(t)|| = \sum_{i \geq 1} \gamma^{i-1} \left| \sum_{j \geq i} (\pi_j - p_j(t)) \right|.$$

Here γ, π_j, p_j (for a simplicity) are not denoted by the sign "S". So the characteristic time $\tau(S_n)$ of the system S_n, understood in the sense of the formula (1.8), satisfies the inequality

$$0 < 1/R_S \leq \tau(S_n) \leq 1/r_S < \infty. \qquad (1.9)$$

Consider now the characteristic time $\tau(\sum_n)$ of the system $\sum_n$, as rule ignoring the index "$\sum$". For this system the quantity

$$a_i = n\mu \left(\rho \left(1 - \frac{d_i}{d_{i-1}} \right) + \min(1, i) - \frac{\min(1, i-1) d_{i-2}}{d_{i-1}} \right), \quad i \geq 1.$$

Choose $d_i = \gamma^i$, $\gamma = 1 + \varepsilon$, $i \geq 0$, $0 < \varepsilon < 1 - \rho$ Then obtain

$$R = nR_\Sigma, \; R_\Sigma = \mu(1 - \rho\varepsilon) < \infty, \qquad (1.10)$$

$$r = n\mu \left(1 - \rho\varepsilon - \frac{1}{1 + \varepsilon} \right) \geq nr_\Sigma,$$

$$r_\Sigma = \frac{\mu\varepsilon}{1 + \varepsilon} (1 - \rho - \rho\varepsilon) > 0. \qquad (1.11)$$

As the formulas (1.10), (1.11) are true so the following inequalities take place for the system $\sum_n$:

$$v_\Sigma e^{-nR_\Sigma t} \leq ||\vec{v}_\Sigma(t)|| \leq v_\Sigma e^{-nr_\Sigma t}, \qquad (1.12)$$

where

$$v_\Sigma = \sum_{i \geq 1} \gamma^{i-1} \sum_{j \geq i} \pi_j < \infty, \quad ||\vec{v}_\Sigma(t)|| = \sum_{i \geq 1} \gamma^{i-1} \left| \sum_{j \geq i} (\pi_j - p_j(t)) \right|.$$

So the characteristic time $\tau(\sum_n)$ of the system $\sum_n$, understood in a sense of the formula (1.12), satisfies the inequality

$$0 < 1/nR_{\Sigma} \le \tau\Big(\sum_n\Big) \le 1/nr_{\Sigma} < \infty. \tag{1.13}$$

The formulas (1.2), (1.3) show that the aggregation of the systems $X_1, \ldots, X_n$ with the type " $\oplus$ " leads to a geometrical rate convergence to zero of the mean stationary queue length for $n \to \infty$. The unification of $X_1, \ldots, X_n$ with the type " $\otimes$ " doesn't influence on the mean stationary queue length.

Controversy the formulas (1.9), (1.13) show that the aggregation of the systems $X_1, \ldots, X_n$ with the type " $\oplus$ " practically doesn't influence on the characteristic time and the aggregation of $X_1, \ldots, X_n$ with the type " $\otimes$ " leads to the convergence of the characteristic time to zero with the rate $\sim 1/n,\ n \to \infty$.

A natural question origins, is it possible to obtain the convergence to zero as for the mean queue so for the characteristic time combining these two types of the aggregation. The positive respond is following.

Suppose that $\Sigma_{n,1}, \ldots, \Sigma_{n,m}$ are m independent queuing systems by the type Σ_n. Consider the system $S_n^m = \Sigma_{n,1} \oplus \ldots \oplus \Sigma_{n,m}$ showed in the figure 3:

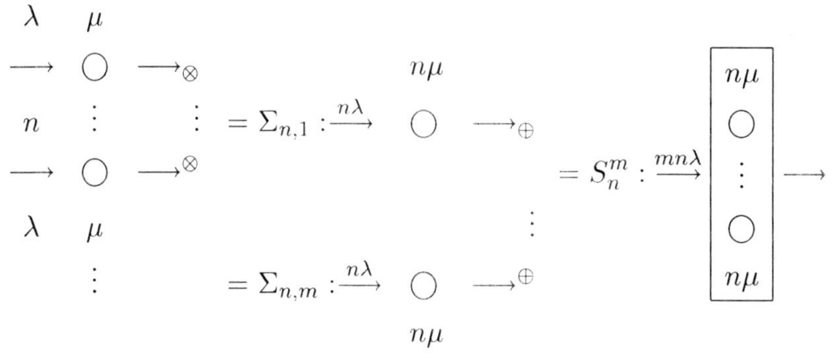

Fig. 3

The formulas (1.2), (1.3), (1.9), (1.13) lead to

$$a(S_n^m) \le Ce^{-\alpha m} \to 0,\ m \to \infty,$$
$$1/nR_{\Sigma} \le \tau(S_n^m) \le 1/nr_{\Sigma} \to 0,\ n \to \infty. \tag{1.14}$$

The queuing system S_n^m is constructed from $N = nm$ the systems with the type $M|M|1|\infty$. So if we take a sufficiently large number N of the initial queuing systems then it is possible to construct. using the aggregations with the types " $\oplus$ ", " $\otimes$ ", the system in which as the characteristic time so the mean stationary queue length are sufficiently small.

Show some formulas which characterize the cooperative effects for the mean stationary waiting time A_n in the system S_n. Denote $\psi = 1 - \rho$, $\overline{C} = e^{1/12}/\sqrt{2\pi}$, $\overline{\psi} = n\psi$.

Theorem 1.1. *1) If $\rho = const$ then*

$$A_n \leq \overline{C} \, \frac{\exp\left(-\frac{n\psi^2}{2-\psi}\right)}{\mu\psi^2 n^{3/2}}. \tag{1.15}$$

2) If $\rho = \rho(n)$, $0 < \psi < a < 1$ then

$$\frac{1}{\mu\overline{\psi}(n)\left(1 + \overline{C}\frac{\overline{\psi}(n)}{\sqrt{n}} \exp\left(\frac{\overline{\psi}^2(n)}{2n(1-a)}\right)\right)} \leq A_n \leq \frac{1}{\mu(1-a)\overline{\psi}(n)}. \tag{1.16}$$

Proof. 1) It is well known [34] that

$$A_n = \frac{\pi_0}{n!} \left(\frac{\lambda n}{\mu}\right)^n \frac{n}{\mu(n - \lambda n/\mu)^2}, \tag{1.17}$$

where

$$\pi_0 = \left(\sum_{k=0}^{n} \frac{1}{k!} \left(\frac{\lambda n}{\mu}\right)^k + \frac{n^n}{n!} \sum_{k=n+1}^{\infty} \left(\frac{\lambda}{\mu}\right)^k\right)^{-1}.$$

Find the upper bound of the quantity A_n. It is clear that

$$\pi_0^{-1} \geq \sum_{k=0}^{n} \frac{1}{k!} \left(\frac{\lambda n}{\mu}\right)^k + \sum_{k>n} \frac{1}{k!} \left(\frac{\lambda n}{\mu}\right)^k = e^{\lambda n/\mu}. \tag{1.18}$$

From (1.18) and the Sterling formula obtain

$$A_n \leq e^{-n\rho} \rho^n \frac{n^{n-1}}{n!\mu(1-\rho)^2} \leq \overline{C} \, \frac{\exp(-n\varphi(\psi))}{\mu\psi^2 n^{3/2}}, \quad n \geq 1, \tag{1.19}$$

where $\varphi(\psi) = -\psi - \ln(1 - \psi)$. The Taylor formula gives that for $\psi \in (0,1)$

$$\varphi(\psi) = -\psi + \psi + \frac{\psi^2}{2} + \frac{\psi^2}{3} + \ldots \geq \frac{\psi^2}{2} \sum_{k\geq 0} \left(\frac{\psi}{2}\right)^k = \frac{\psi^2}{2-\psi}. \tag{1.20}$$

Combining the inequalities (1.20) and (1.19) obtain (1.15).

2) As the inequality

$$\pi_0^{-1} \geq \frac{n^n}{n!} \sum_{k>n} \rho^k = \rho \frac{n^n \rho^n}{n! \psi}$$

is true so

$$A_n \leq \frac{1}{\mu(1-a)\overline{\overline{\psi}}} \ .$$

Find now the low bound of the quantity A_n. As

$$\pi_0^{-1} = \sum_{k=0}^{n} \frac{\rho^k}{k!} + \frac{n^n}{n!} \sum_{k>n} \rho^k \leq e^{n\rho} + \frac{n^n(1-\psi)^n}{n!\psi}$$

so from the Sterling formula obtain

$$A_n \geq \frac{1}{\mu \overline{\psi}(n)(1+\beta(n))} \ ,$$

where

$$\beta(n) = \overline{C}\psi \frac{e^{-n\psi}}{(1-\psi)^n} \sqrt{n} \leq \overline{C}\psi \sqrt{n} \exp\left(n \frac{\psi^2}{2(1-a)}\right) =$$

$$= \frac{\overline{C}\,\overline{\psi}}{\sqrt{n}} \exp\left(\frac{\overline{\psi}^2}{2n(1-a)}\right) .$$

Two-sided bounds (1.16) are proved.

Corollary 1.1. *Suppose that $\psi = \psi(n) = 1 - \rho(n) \sim n^{-\alpha}$ then for $\alpha < 1$ $A_n \to 0$, $n \to \infty$ and for $\alpha > 1$ $A_n \to \infty$, $n \to \infty$.*

Remark 1. *A characteristic feature of a main part of known queuing systems is an increasing of the mean stationary waiting time and the mean stationary queue length for ultimate meanings of the load coefficient. In an our consideration if the number n of the systems $M|M|1|\infty$, aggregated into the system S_n, tends to the infinity then new phenomenon analogous to the phase transition in physical systems is recognized.*

Nonidentical servers

In the previous subsection a few bounds of the cooperative effects for the aggregation of the identical oneserver queuing systems into multiserver queuing system have been considered. In this subsection twoserver queuing system S_2 combined from the systems $M|M|1|\infty$ with the identical load coefficients but different serving intensities is considered. The load coefficients of the separate servers and the stationary probabilities of the customer to be served at one of these servers are investigated.

Denote λ', λ'' the input intensities and μ', μ'' the serving intensities of the initial one-server queuing systems. Then the input intensity λ in the aggregated system S_2 equals to $\lambda' + \lambda''$. Denote $\mu = = \mu_1 + \mu_2$ and suppose that the load coefficient $\rho = \lambda'/\mu' = \lambda''/\mu'' == \lambda/\mu$.

To describe a functioning of the aggregated twoserver queueing system consider the discrete Markov process $i(t)$, $t \geq 0$, with the set state $\{0, 1', 1'', 2, 3, \ldots\}$. Here the equalities $i(t) = 0$, $i(t) = 2$, $i(t) = 3, \ldots$ designate that in the aggregated twoserver queueing system at the moment t there are $0, 2, 3, \ldots$ customers. The equalities $i(t) = 1'$, $i(t) = 1''$ designate that at the moment t there is a single customer which is served at the server with the intensity μ', μ'', accordingly. The nonzero transition intensities of the process $i(t)$ are defined as follows:

$$L(0, 1') = \lambda', \ L(0, 1'') = \lambda'', \ L(1', 0) = \mu', \ L(1'', 0) = \mu'',$$

$$L(1', 2) = \lambda, \ L(1'', 2) = \lambda, \ L(2, 1') = \mu'', \ L(2, 1'') = \mu',$$

$$L(i, i - 1) = \mu, \ i \geq 3, \ L(i, i + 1) = \lambda, \ i \geq 2.$$

It is simple to prove [29, 37] that for $\rho < 1$ the Markov process $i(t)$ has the limit probabilities $p(0), p(1'), p(1''), p(2), p(3), \ldots$, satisfying the following system of the linear algebraic equations

$$
\begin{cases}
\lambda p(0) = \mu' p(1') + \mu'' p(1''), \\
(\lambda + \mu') p(1') = \lambda' p(0) + \mu'' p(2), \\
(\lambda + \mu'') p(1'') = \lambda'' p(0) + \mu' p(2), \\
(\lambda + \mu) p(2) = \lambda(p(1') + p(1'')) + \mu p(3), \\
(\lambda + \mu) p(i) = \lambda p(i - 1) + \mu p(i + 1), \quad i = 3, 4, \ldots, \\
p(0) + p(1') + p(1'') + p(2) + p(3) + \ldots = 1.
\end{cases}
\tag{1.21}
$$

Denote $\alpha = \mu'/\mu = \lambda'/\lambda$, $\beta = \alpha(1 - \alpha)$ and analyze how do the quantities $A = \pi(1')/\pi(1'')$, $B = p(0) + p(1') + p(1'')$ depend on the parameters α (or β) and ρ. The quantity A characterizes the difference between stationary probabilities

$\pi(1') = p(1') + p(2) + p(3) + \ldots,$

$\pi(1'') = p(1'')+p(2)+p(3)+\ldots$ These probabilities characterize the different servers load. The quantity B characterizes the stationary probability that an input customer does not wait a serving.

Denote $p(1) = p(1') + p(1'')$ then from the system (1.21) obtain

$$\lambda p(1)=\mu p(2), \quad (\lambda+\mu)p(i)=\lambda p(i-1)+\mu p(i+1), \quad i \geq 2. \tag{1.22}$$

The equalities (1.22) lead to the formula

$$p(i) = \rho^{i-1}p(1), \quad i = 2, 3, \ldots \tag{1.23}$$

Accordingly to the formula (1.23) and to the first equation of the system (1.21) obtain

$$A = \frac{(1 - \alpha)(\alpha + \rho)}{\alpha(1 - \alpha + \rho)}, \quad B = \frac{\beta(1 + 4\rho)(1 - \rho) + \rho^2(1 - \rho)}{\beta[(1 + 2\rho)(1 - \rho) + 2\rho] + \rho^2}. \tag{1.24}$$

The formula (1.24) leads to the equivalence of the inequalities

$$A < 1 \iff \alpha > \frac{1}{2}. \tag{1.25}$$

For fixed α, $1/2 < \alpha \leq 1$, there is the monotone convergence $A \downarrow (1-\alpha^2)/(\alpha(2-\alpha))$ if $\rho \uparrow 1$. For fixed ρ, $0 < \rho < 1$, there is the convergence $A \to 0$ if $\alpha \to 1$. So in the aggregated two-server queuing system a faster server has a smaller load probability. This effect enforces if the coefficient ρ increases and if the coefficient α (characterizing relative intensity of the faster server) tends to 1. If $\beta = \alpha(1 - \alpha)$ increases, $0 \leq \beta \leq 1/4$, then the coefficient B monotonically increases too. So the stationary probability B of a waiting absence increases if the difference between the servers intensities decreases.

Fix ρ, $0 < \rho < 1$, and consider the dependencies $A = A(\varepsilon)$, $B = B(\varepsilon)$, $\varepsilon = \alpha - 1/2$, for $\varepsilon \to 0$. The formulas (1.24) allow to construct the following Taylor rows:

$$A(\varepsilon) = 1 - \varepsilon a + o_1(\varepsilon), \quad B(\varepsilon) = B(0) - \varepsilon^2 b + o_2(\varepsilon),$$
$$\lim_{\varepsilon \to 0} o_1(\varepsilon)/\varepsilon = 0, \quad \lim_{\varepsilon \to 0} o_2(\varepsilon)/\varepsilon = 0, \quad B(0), \ a, \ b > 0. \tag{1.26}$$

So a small (with the order ε) difference between servers intensities leads to the decreasing of A (with the order ε) and to the decreasing of B (with the order ε^2). As a result a small difference between servers intensities influences stronger on internal than on external characteristics of the aggregated system.

1.2. Product form stationary distributions of Markov processes

In this subsection the theorem which allows to calculate the stationary distributions of multi-component Markov processes in the product form is proved. This result is obtained for a special form of transition intensities of N-dimension Markov process via transition intensities of N one-dimension Markov processes. Considered multi-dimension Markov processes possess the ergodicity and the invariance properties.

The product theorem allow to calculate stationary distributions of queueing networks in a random environment. Stationary distribution of mobile telephone system with a random input intensity and a controlled load coefficient is calculated.

Product theorem

It is known that the homogenous Markov process $x(t)$, $t \geq 0$, with the discrete state set X and the transition intensities λ_{ij}, $i, j \in X$, is ergodic if

(A) the system of the equations

$$u_j \sum_{i \in X} \lambda_{ji} = \sum_{i \in X} u_i \lambda_{ij}, \; j \in X, \tag{1.27}$$

has at least one solution $\{u_i\}$ so that $0 < \sum_{i \in X} |u_i| < \infty$,

(B) all states of the process $x(t)$, $t \geq 0$, are communicable:

$$\forall i, i^* \in X \; \exists i_1, ..., i_r \in X : \; \lambda_{ii_1} > 0, \; \lambda_{i_1 i_2} > 0, ..., \lambda_{i_r i^*} > 0. \tag{1.28}$$

(C) the following regularity condition is true:

$$\exists \Lambda < \infty : \; \forall j \in X \sum_{i \in X} \lambda_{ji} < \Lambda. \tag{1.29}$$

Then [34], [45] the stationary distribution of this process coincides with its limit (ergodic) distribution and is defined uniquely from the system (1.27) and the normalization condition. If the state set is finite and all process states are communicable then the conditions **(A)** and **(C)** are true too.

Consider the set of the homogenous Markov processes $\{x_k(t), k = \overline{1, N}\}$ with the discrete state sets X_k, the transition intensities $\lambda_k(i_k, j_k)$, $i_k, j_k \in X_k$, which satisfy the conditions **(A)** - **(C)**. Denote

$$I = \{1, 2, ..., N\}, \; c_0 = const \geq 0,$$

$$\Omega_s = \{\{k_1, ..., k_s\} \subset I, \ k_i \neq k_j, \ i, j = \overline{1, s}, \}, \ s = \overline{1, N-1},$$

$$X = \otimes_{i \in I} X_i, \ \ X^{(k_1,...,k_s)} = \otimes_{i \in I \setminus \{k_1,...,k_s\}} X_i, \ \ \vec{i} = (i_1, ..., i_N),$$

$$c_{k_1,...,k_s} : \otimes_{i \in \{k_1,...,k_s\}} X_i \rightarrow R_+ \cup \{0\}, \ \ \{k_1, ..., k_s\} \in \Omega_s.$$

Construct the Markov process $(x'_1(t), ..., x'_N(t))$ the with state set X and the transition intensities

$$L(\vec{i}, \vec{j}) = \begin{cases} c_{k_1,...,k_s}(i_{k_1}, ..., i_{k_s}) \displaystyle\prod_{k \in I \setminus \{k_1,...,k_s\}} \lambda_k(i_k, j_k), \ j_{k_l} = i_{k_l}, l = \overline{1, s}, \\[2mm] c_0 \displaystyle\prod_{k \in I} \lambda_k(i_k, j_k), \ j_l \neq i_l, \ l = \overline{1, N}, \\[2mm] 0, \ \vec{i} = \vec{j}. \end{cases}$$

Theorem 1.2. *If*

$$\forall \{k_1, ..., k_{N-1}\} \in \Omega_{N-1} \ \ c_{k_1,...,k_{N-1}}(i_{k_1}, ..., i_{k_{N-1}}) > 0, \tag{1.30}$$

$$\max_{s=\overline{1, N-1}} \max_{\{k_1,...,k_s\} \in \Omega_s} c_{k_1,...,k_s}(i_{k_1}, ..., i_{k_s}) < C < \infty, \tag{1.31}$$

then the process $(x'_1(t), ..., x'_N(t))$ *is ergodic and its stationary distribution has the product form:*

$$P_{\vec{i}} = \prod_{k \in I} \pi_{i_k}(k), \tag{1.32}$$

where $\pi_{i_k}(k)$, $i_k \in X_k$, *is the stationary distribution of the process* $x_k(t)$.

Proof. Denote

$$\lambda^{(k_1,...,k_s)}(\vec{i}, \vec{j}) = c_{k_1,...,k_s}(i_{k_1}, ..., i_{k_s}) I(j_{k_l} = i_{k_l}, \ l = \overline{1, s}) \prod_{k \in I \setminus \{k_1,...,k_s\}} \lambda_k(i_k, j_k),$$

$$\{k_1, ..., k_s\} \in \Omega_s, \ s = \overline{1, N-1},$$

$$\lambda^{(0)}(\vec{i}, \vec{j}) = c_0 I(j_l \neq i_l, \ l = \overline{1, N}) \prod_{k \in I} \lambda_k(i_k, j_k).$$

The proof of the condition **(A)** for the process $(x'_1(t), ..., x'_N(t))$ is made in two steps. The first step:

$$P_{\vec{i}} \sum_{\vec{j} \in X} \lambda^{(k_1,...,k_s)}(\vec{i}, \vec{j}) =$$

$$= \prod_{k \in I} \pi_{i_k}(k) \sum_{\vec{j} \in X} c_{k_1,\ldots,k_s}(i_{k_1},\ldots,i_{k_s}) I(j_{k_l} = i_{k_l},\ i = \overline{1,s}) \prod_{k \in I \setminus \{k_1,\ldots,k_s\}} \lambda_k(i_k, j_k) =$$

$$= \prod_{k \in I} \pi_{i_k}(k) c_{k_1,\ldots,k_s}(i_{k_1},\ldots,i_{k_s}) \sum_{\substack{(j_{t_1},\ldots,j_{t_{N-s}}) \in X^{(t_1,\ldots,t_{N-s})} \\ \{t_1,\ldots,t_{N-s}\} \in I \setminus \{k_1,\ldots,k_s\}}} \prod_{k \in I \setminus \{k_1,\ldots,k_s\}} \lambda_k(i_k, j_k) =$$

$$= \prod_{k \in I} \pi_{i_k}(k) c_{k_1,\ldots,k_s}(i_{k_1},\ldots,i_{k_s}) \prod_{k \in I \setminus \{k_1,\ldots,k_s\}} \sum_{j_k \in X_k} \lambda_k(i_k, j_k) =$$

$$= \prod_{k \in \{k_1,\ldots,k_s\}} \pi_{i_k}(k) c_{k_1,\ldots,k_s}(i_{k_1},\ldots,i_{k_s}) \prod_{k \in I \setminus \{k_1,\ldots,k_s\}} \pi_{i_k}(k) \sum_{j_k \in X_k} \lambda_k(i_k, j_k) =$$

$$= \prod_{k \in \{k_1,\ldots,k_s\}} \pi_{i_k}(k) c_{k_1,\ldots,k_s}(i_{k_1},\ldots,i_{k_s}) \prod_{k \in I \setminus \{k_1,\ldots,k_s\}} \sum_{j_k \in X_k} \pi_{j_k}(k) \lambda_k(j_k, i_k) =$$

$$= \sum_{\substack{(j_{t_1},\ldots,j_{t_{N-s}}) \in X^{(t_1,\ldots,t_{N-s})} \\ \{t_1,\ldots,t_{N-s}\} \in I \setminus \{k_1,\ldots,k_s\}}} \prod_{k \in I \setminus \{k_1,\ldots,k_s\}} c_{k_1,\ldots,k_s}(i_{k_1},\ldots,i_{k_s}) \times$$

$$\times \prod_{l \in \{k_1,\ldots,k_s\}} \pi_{i_l}(l) \pi_{j_k}(k) \lambda_k(j_k, i_k) =$$

$$= \sum_{\vec{j} \in X} \prod_{k \in I} \pi_{j_k}(k) c_{k_1,\ldots,k_s}(i_{k_1},\ldots,i_{k_s}) I(i_{k_l} = j_{k_l},\ i = \overline{1,s}) \times$$

$$\times \prod_{k \in I \setminus \{k_1,\ldots,k_s\}} \lambda_k(j_k, i_k) = \sum_{\vec{j} \in X} P_{\vec{j}} \lambda^{(k_1,\ldots,k_s)}(\vec{j}, \vec{i}),$$

for $\{k_1,\ldots,k_s\} \in \Omega_s,\ s = \overline{1, N-1}$. Analogously check that

$$P_{\vec{i}} \sum_{\vec{j} \in X} \lambda^{(0)}(\vec{i}, \vec{j}) = \sum_{\vec{j} \in X} P_{\vec{j}} \lambda^{(0)}(\vec{j}, \vec{i}).$$

The second step:

$$P_{\vec{i}} \sum_{\vec{j} \in X} L(\vec{i}, \vec{j}) = P_{\vec{i}} \sum_{\vec{j} \in X} \left(\sum_{s=1}^{N-1} \sum_{\{k_1,\ldots,k_s\} \in \Omega_s} \lambda^{(k_1,\ldots,k_s)}(\vec{i}, \vec{j}) + \lambda^{(0)}(\vec{i}, \vec{j}) \right) =$$

$$= \sum_{s=1}^{N-1} \sum_{\{k_1,\ldots,k_s\} \in \Omega_s} P_{\vec{i}} \sum_{\vec{j} \in X} \lambda^{(k_1,\ldots,k_s)}(\vec{i}, \vec{j}) + P_{\vec{i}} \sum_{\vec{j} \in X} \lambda^{(0)}(\vec{i}, \vec{j}) -$$

$$= \sum_{s=1}^{N-1} \sum_{\{k_1,\ldots,k_s\} \in \Omega_s} \sum_{\vec{j} \in X} P_{\vec{j}} \lambda^{(k_1,\ldots,k_s)}(\vec{j}, \vec{i}) + \sum_{\vec{j} \in X} P_{\vec{j}} \lambda^{(0)}(\vec{j}, \vec{i}) =$$

$$= \sum_{\vec{j} \in X} P_{\vec{j}} \left(\sum_{s=1}^{N-1} \sum_{\{k_1,\ldots,k_s\} \in \Omega_s} P_{\vec{j}} \lambda^{(k_1,\ldots,k_s)}(\vec{j},\vec{i}) + \lambda^{(0)}(\vec{j},\vec{i}) \right) = \sum_{\vec{j} \in X} P_{\vec{j}} L(\vec{j},\vec{i}).$$

It means that the condition **(A)** is true and

$$\sum_{\vec{i} \in X} P_{\vec{i}} = \prod_{k \in I} \sum_{i_k \in X_k} \pi_{i_k}(k) = 1.$$

As for each k, $k = \overline{1,N}$, all states of the set X_k are communicable:

$$\forall \, i_k, i_k^* \in X_k \, \exists \, i_k^1, \ldots, i_k^{r_k} \in X_k \, : \, \lambda_k(i_k, i_k^1) > 0, \ldots, \lambda_k(i_k^{r_k}, i_k^*) > 0,$$

and the conditions (1.30) are true so for $\forall \vec{i}, \vec{i^*} \in X$

$$L((i_1, i_2, \ldots, i_N), (i_1^1, i_2, \ldots, i_N)) > 0, \, L((i_1^1, i_2, \ldots, i_N), (i_1^2, i_2, \ldots, i_N)) > 0, \ldots,$$

$$L((i_1^{r_1}, i_2, \ldots, i_N), (i_1^*, i_2, \ldots, i_N)) > 0, \, L((i_1^*, i_2, \ldots, i_N), (i_1^*, i_2^1, \ldots, i_N)) > 0, \ldots,$$

$$L((i_1^*, i_2^*, \ldots, i_{N-1}^*, i_N^{r_N}), (i_1^*, i_2^*, \ldots, i_N^*)) > 0.$$

Consequently the condition **(B)** is true for the considered process.

Denote by Λ_k, $k = 1, \ldots, N$, the constants, found from the regularity condition for the processes $x_k(t)$, $k = 1, \ldots, N$. Then from the condition (1.31) obtain

$$\sum_{\vec{j} \in X} \lambda^{(k_1,\ldots,k_s)}(\vec{i},\vec{j}) < C \sum_{\substack{(j_{t_1},\ldots,j_{t_{N-s}}) \in X^{(t_1,\ldots,t_{N-s})} \\ \{t_1,\ldots,t_{N-s}\} \in I \backslash \{k_1,\ldots,k_s\}}} \prod_{k \in I \backslash \{k_1,\ldots,k_s\}} \lambda_k(i_k, j_k) =$$

$$= C \prod_{k \in I \backslash \{k_1,\ldots,k_s\}} \sum_{j_k \in X_k} \lambda_k(i_k, j_k) < C \prod_{k \in I \backslash \{k_1,\ldots,k_s\}} \Lambda_k < \infty,$$

$$\{k_1, \ldots, k_s\} \in \Omega_s, \, s = \overline{1, N-1}$$

and

$$\sum_{\vec{i} \in X} \lambda^{(0)}(\vec{i},\vec{j}) < C \prod_{k \in I} \lambda_k(i_k, j_k) < C \prod_{k \in I} \Lambda_k.$$

So

$$\sum_{\vec{j} \in X} L(\vec{i},\vec{j}) < C \left(\sum_{s=1}^{N-1} \sum_{\{k_1,\ldots,k_s\} \in \Omega_s} \prod_{k \in I \backslash \{k_1,\ldots,k_s\}} \Lambda_k + \prod_{k \in I} \Lambda_k \right) < \infty.$$

Consequently the condition **(C)** is true for the process $(x_1'(t), \ldots, x_N'(t))$ too.

Corollary 1.2. *The limit distribution of the process $x'_k(t)$, considered as the k-th component of the process $(x'_1(t), ..., x'_N(t))$, coincides with the limit distribution of the process $x_k(t)$, $k = \overline{1, N}$.*

Proof.

$$\lim_{t \to \infty} P(x'(t) = i_k) = \lim_{t \to \infty} \sum_{i_l \in X_l, \, l \neq k} P_{\vec{i}} = \pi_{i_k}(k) = P(x(t) = i_k).$$

Remark 2. *It is possible to replace the condition (1.30) in the theorem 1.2 by the condition:*

$$c_0 > 0 \ or \ \forall \vec{i}, \vec{i^*} \in X \ \exists \, r : r_1 = r_2 = ... = r_N = r,$$

*where $r_1, ..., r_N$ are defined from the condition **(B)** for the processes $x_1(t), ..., x_N(t)$ or in general case the condition **(B)** is true for the process $(x'_1(t), ..., x'_N(t))$.*

Remark 3. *The theorem 1.2 generalizes the result of F. Kelly [43], which allows to use reversibility property for a calculation of the stationary distribution only in the case, when $c_0 = 0$ and $c_{k_1,...,k_s}(i_{k_1}, ..., i_{k_s}) = 0$ for $\forall \{k_1, ..., k_s\} \in \Omega_s$, $s = \overline{1, \ N-2}$.*

Distribution of customers numbers in nodes

of mobile telephone network

Consider mobile telephone network, which includes two stations numbered $1, 2$ and consisted of single servers, and two switching centers numbered $3, 4$ and consisted of infinite number of identical servers. The flow of input customers is controlled in the following way: the input flow is reared and may have the intensities λ^* or λ^{**}. The stations and the switching centers are adopted to the input intensities by a varying of their serving intensities. Customers, moving from one station to another station, pass through the switching centers arranged between them.

Suppose that the input flow of the network is Poisson with the intensity $\lambda(t)$, where $\lambda(t)$ is the homogenous Markov process with the set of the communicable states $\Lambda = \{\lambda_1, ..., \lambda_h\}$ and the transition intensities $\gamma(\lambda_i, \lambda_j)$, $\lambda_i, \lambda_j \in \Lambda$. If $\lambda^* \leq \lambda_i < \lambda^{**}$ then the network input flow has the intensity λ^* and each server in the i-th node has the intensity μ_i. If $\lambda_i \geq \lambda^{**}$ then the network input flow has the intensity λ^{**} and each server in the $i-$th station has intensity $\mu_i \lambda^{**}/\lambda^*$. The customers motion is defined by the route matrix $||\theta_{ij}||_{i,j=\overline{0,2}}$, where θ_{ij} is the transition probability to reach the $j-$th node after the serving in the i-th node, $\theta_{00} = 0$.

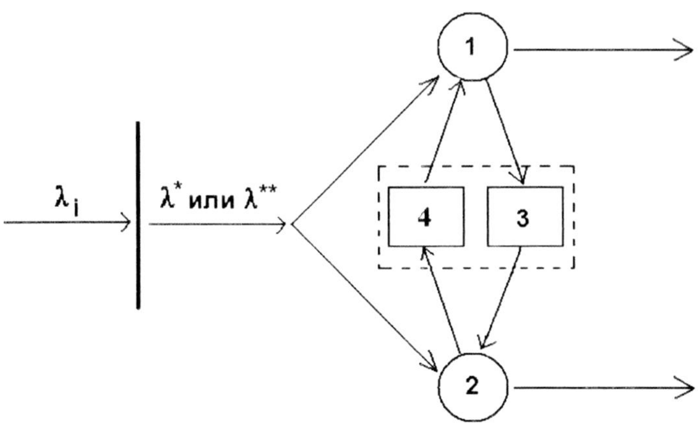

Fig. 4

The constructed mobile telephone model in the case of constant input intensity λ^* is so called $BCMP$-network [62]. The functioning of such the network is described by the Markov process $x(t)$ with the state set $D = \{\vec{n} = (n_1, ..., n_4), \; n_i \geq 0\}$, where n_i is the number of customers in the i-th node and with the transition intensities

$$\beta_{\vec{n},\vec{n}'} = \begin{cases} \lambda^* \theta_{0i}, \; \vec{n}' = \vec{n} + \vec{N}_i, \; i = 1, 2, \\ \mu_i n_i, \; \vec{n}' = \vec{n} + \vec{N}_j - \vec{N}_i, \; (i,j) \in \{(3,2), (4,1)\}, \; n_i > 0, \\ \mu_1 \theta_{12}, \; \vec{n}' = \vec{n} + \vec{N}_3 - \vec{N}_1, \; n_1 > 0, \\ \mu_2 \theta_{21}, \; \vec{n}' = \vec{n} + \vec{N}_4 - \vec{N}_2, \; n_2 > 0, \\ \mu_i \theta_{i0}, \; \vec{n}' = \vec{n} - \vec{N}_i, \; i = 1, 2, \; n_i > 0. \end{cases}$$

Here $\vec{N}_i$ is the vector with the i-th component equal to 1 and the other components equal to zero. If $e_i < \mu_i$, $i = 1, 2$, where e_i, $i = \overline{1, 4}$, are defined from the system of the equalities

$$\begin{cases} e_3 = e_1 \theta_{12}, \\ e_4 = e_2 \theta_{21}, \\ \lambda^* \theta_{01} + e_2 \theta_{21} = e_1, \\ \lambda^* \theta_{02} + e_1 \theta_{12} = e_2, \end{cases}$$

then the stationary distribution of the process $x(t)$ is calculated as follows

$$P_{\vec{n}} = \prod_{i=1}^{2} \frac{\mu_i}{\mu_i - e_i} \left(\frac{e_i}{\mu_i}\right)^{n_i} \prod_{i=3}^{4} \frac{\exp\left(-e_i/\mu_i\right)}{n_i!} \left(\frac{e_i}{\mu_i}\right)^{n_i}. \tag{1.33}$$

If the input intensity $\lambda(t)$ is varying then the constructed model is described by the two-dimensional Markov process with the state set $\Lambda \times D$ and the transition intensities

$$L((\lambda_i, \vec{n}), (\lambda_j, \vec{n}')) = \begin{cases} \beta_{\vec{n}, \vec{n}'}, & j = i, \ \lambda^* \leq \lambda_i < \lambda^{**}, \\ \frac{\lambda^{**}}{\lambda^*} \beta_{\vec{n}, \vec{n}'}, & j = i, \ \lambda_i \geq \lambda^{**}, \\ \gamma(\lambda_i, \lambda_j), & \vec{n}' = \vec{n}. \end{cases}$$

Accordingly to the theorem 1.2 the stationary distribution of this process has the form:

$$\mathbf{P}_{(\lambda_i, \vec{n})} = \Pi_i P_{\vec{n}}, \ i = \overline{1, h}, \ \vec{n} \in D,$$

where Π_i, $i = \overline{1, h}$, is the stationary distribution of the process $\lambda(t)$. The ergodicity of the considered process takes place accordingly to the sufficient conditions (**A**) - (**C**).

Remark 4. *The switching centers 3, 4 may be aggregated into the common node with a conservation of an initial order of a serving and a routing of the customers. Then in the formula (1.33) the product of Poisson distributions with the parameters e_3/μ_3, e_4/μ_4 are to be replaced by their conjuncture which is Poisson distribution with the parameter $e_3/\mu_3 + e_4/\mu_4$.*

1.3. Product theorems for queueing networks with switching centers

The product formula of stationary distribution calculation, obtained in the previous subsection for mobile telephone network, is generalized here onto general queueing networks. Considered networks work so that a customer, moving from one station to another station, passes through some switching center. The switching center is a queueing system with an infinite number of servers, which have exponential distribution of serving times. This center does not change the customers routings. So the classical product theorems [62] can not be applied to the networks with switching centers. In this subsection a special algorithm of the stationary distribution calculation is constructed. It is based on the classical product theorems and the additive property of Poisson distribution.

Opened networks

The network G is classical Jackson network (partial case of $BCMP$-network) [62] if it has Poisson input flow with the intensity λ and consists of m multi-server queueing systems with the exponential distributions of the serving times. Denote r_i, μ_i the number of servers and the intensity of the serving in the server of the i−th node correspondingly. An external source of the input flow may be numbered by 0. A dynamics of the customers motion is defined by the route matrix $\Theta = (\theta_{ij})_{i,j=0}^{m}$, where θ_{ij} is the probability of the transition from the i−th node to the j−th node, $\theta_{00} = 0$. Suppose that the route matrix is indivisible:

$$\forall i, j \in \{0, 1, ..., m\} \; \exists \; i_1, \, i_2, \ldots i_r \in \{1, ..., m\} \; :$$

$$\theta_{ii_1} > 0, \; \theta_{i_1 i_2} > 0, \ldots \theta_{i_r j} > 0.$$

Then the vector $(\lambda, \overline{\lambda}_1, \overline{\lambda}_2, ..., \overline{\lambda}_m)$ is the single solution of the system

$$(\lambda, \overline{\lambda}_1, \overline{\lambda}_2, ..., \overline{\lambda}_m) = (\lambda, \overline{\lambda}_1, \overline{\lambda}_2, ..., \overline{\lambda}_m)\Theta.$$

In an isolated regime the i-th node of the network is r_i-server queueing system with Poisson input flow, which intensity equals to $\overline{\lambda}_i$. A current number of the customers in this network is described by the birth and death process. If

$$\rho_i = \frac{\overline{\lambda}_i}{r_i \mu_i} < 1, \tag{1.34}$$

then [34] this process is ergodic and its stationary distribution is following

$$\pi_i(j) = \frac{a_i(j)}{1 + \sum_{s>0} a_i(s)}, \; j \geq 0, \;\; a_i(s) = \prod_{k=1}^{s} \frac{\overline{\lambda}_i}{\min(k, \, r_i)\mu_i}, \tag{1.35}$$

where $s > 0$, $a_i(0) = 1$.

A functioning of a complete network (the current numbers of the customers in the nodes) is described by the discrete Markov process $z(t) = (n_1(t), \ldots, n_m(t))$ with state set $D = \{(n_1, ..., n_m) \; : \; n_1 \geq 0, ..., n_m \geq 0\}$. If the formulas (1.34) are true for $i = 1, 2, \ldots, m$ then the Markov process $z(t) = (n_1(t), \ldots, n_m(t))$ is ergodic [34] and its stationary distribution $P(n_1, ..., n_m)$ is following [62]:

$$P(n_1, ..., n_m) = \prod_{i=1}^{m} \pi_i(n_i), \;\; (n_1, ..., n_m) \in D. \tag{1.36}$$

On the edge (i, j), $1 \leq i, j \leq m$, of the network G put the switching center (the infinite-server queuing system with exponentially distributed serving times, which intensities equal to μ_{ij}). The switching centers do not change the routings of the customers. Denote this network by G_1.

If

$$\pi_{ij}(n_{ij}) = \frac{e^{-a_{ij}}}{n_{ij}!}a_{ij}^{n_{ij}}, \quad a_{ij} = \frac{\overline{\lambda}_i\theta_{ij}}{\mu_{ij}}, \quad j \neq i, \ 1 \leq i, \ j \leq m, \tag{1.37}$$

then the stationary distribution of the process $z_1(t) = (n_i(t), \ n_{ij}(t),$ $j \neq i, \ 1 \leq i, \ j \leq m)$, describing the network G_1 (here $n_{ij}(t)$ is the number of the customers in the switching center arranged on the edge (i,j) at the moment t) is calculated by the formula [62]:

$$P_1(n_i, \ n_{ij}, \ j \neq i, \ 1 \leq i, \ j \leq m) =$$

$$= P(n_1, ..., n_m) \prod_{1 \leq i,j \leq m} \pi_{ij}(n_{ij}), \quad n_i \geq 0, \ n_{ij} \geq 0. \tag{1.38}$$

as the network G_1 is $BCMP$-network.

Divide the edge set (i, j), $1 \leq i, j \leq m$, into the nonintersecting subsets $K_1, \ldots, K_r$ and denote

$$A_{K_s} = \sum_{(i,j) \in K_s} a_{ij}, \quad 1 \leq s \leq r.$$

Aggregate the switching centers arranged on the edges $(i, j) \in K_s$, $1 \leq s \leq r$ into the common switching centers with the infinite number of servers and the serving intensities depending on the customers routings. Denote such the network by G_2.

The network G_2 is described by the random process $z_2(t) = (n_i(t), \ 1 \leq i \leq m, \ N_s(t), \ 1 \leq s \leq r)$, where $N_s(t)$ is the number of the customers in the s-th common switching center at the moment t. It is clear that

$$N_s(t) = \sum_{(i,j) \in K_s} n_{ij}(t), \quad 1 \leq s \leq r, \tag{1.39}$$

and the process $z_2(t)$ is not Markov one.

Theorem 1.3. *The stationary distribution of the process $z_2(t)$ is following*

$$P_2(n_i, \ 1 \leq i \leq m, \ N_s, \ 1 \leq s \leq r) = P(n_1, ..., n_m) \prod_{s=1}^{r} \Pi_s(N_s), \tag{1.40}$$

$$\Pi_s(N_s) = \frac{e^{-A_{K_s}}}{N_s!}A_{K_s}^{N_s}, \quad n_i \geq 0, \ N_s \geq 0.$$

Proof. The theorem 1.3 proof is based on the formulas (1.37) – (1.39) and on the additive property of Poisson distribution: if the independent random variables x, y have Poisson distributions with the parameters a, b then their sum $x + y$ has Poisson distribution with the parameter $a + b$.

Remark 5. *This algorithm may be easily spread onto a case when each edge of the network G is replaced by an opened Jackson network with nodes, containing the infinite numbers of their servers.*

Closed networks

The closed network G' (Gordon-Newel network) differs from the opened network by the condition that constant number M of the customers circulate in this network and there is not the input flow. The motion of the customers in the network G' are described by the route matrix $\Theta' = (\theta_{ij})_{i,j=1}^{m}$, where θ_{ij} is the probability of the transition from the $i-$th node to the $j-$th node and the matrix Θ' is indivisible:

$$\forall i, j \in \{1, ..., m\} \; \exists \, i_1, i_2, \ldots i_r \in \{1, ..., m\} : \; \theta_{ii_1} > 0, \theta_{i_1 i_2} > 0, \ldots, \theta_{i_r j} > 0.$$

Then for each $B > 0$ the solution $(\overline{\lambda}_1, \overline{\lambda}_2, ...\overline{\lambda}_m)$ of the system

$$\sum_{i=1}^{m} \overline{\lambda}_i = B, \;\; (\overline{\lambda}_1, \overline{\lambda}_2, ..., \overline{\lambda}_m) = (\overline{\lambda}_1, \overline{\lambda}_2, ..., \overline{\lambda}_m)\Theta'$$

exists and is unique.

The functioning of this network (the current numbers of the customers in the nodes) is described by the discrete Markov process $z'(t) = (n_1(t), \ldots, n_m(t))$ with the state set $D' = \{(n_1, ..., n_m) : n_1, ..., n_m \geq 0, \sum_{i=1}^{m} n_i = M\}$. The process $z'(t)$ is ergodic [34] and its stationary distribution is calculated by the formula [62]:

$$P'(n_1. ...n_m) = C^{-1} \prod_{i=1}^{m} a_i(n_i), \; (n_1, ..., n_m) \in D', \tag{1.41}$$

where

$$C = \sum_{(n_1,...,n_m) \in D'} \prod_{i=1}^{m} a_i(n_i),$$

$a_i(n_i)$ is calculated by the formulas from (1.35).

Put the switching center on each edge (i, j), $1 \leq i, j \leq m$, of the network G'. Denote the constructed network by G'_1. The stationary distribution of the process $z'_1(t) = (n_i(t), n_{ij}(t), j \neq i, 1 \leq i, j \leq m)$, describing the network G'_1, is calculated by the formula [62]:

$$P'_1(n_i, n_{ij}, 1 \leq i, j \leq m) =$$

$$= C_1^{-1} P'(n_1, ..., n_m) \prod_{1 \leq i, j \leq m} \pi_{ij}(n_{ij}), \tag{1.42}$$

$$n_i \geq 0, \quad n_{ij} \geq 0, \quad 0 \leq i, j \leq m, \quad \sum_{i=1}^{m} n_i + \sum_{i,j=1}^{m} n_{ij} = M,$$

$$C_1 = \sum_{n_i \geq 0,\, n_{ij} \geq 0,\, \sum_{i=1}^{m} n_i + \sum_{i,j=1}^{m} n_{ij} = M} P'(n_1, ..., n_m) \prod_{1 \leq i, j \leq m} \pi_{ij}(n_{ij}),$$

$\pi_{ij}(n_{ij})$ are calculated by the formulas (1.37).

Analogously divide now the edge set (i, j), $1 \leq i, j \leq m$, into the nonintersecting subsets $K_1, \ldots, K_r$ and aggregate the switching centers arranged on the edges $(i, j) \in K_s$, $1 \leq s \leq r$, into the common switching centers. Denote such the network by G'_2.

The functioning of the network G'_2 is described by the random process $z'_2(t) = (n_i(t), 1 \leq i \leq m, N_s(t), 1 \leq s \leq r)$, where $N_s(t)$ is the number of the customers in the s-th switching center K_s at the moment t,

$$N_s(t) = \sum_{(i,j) \in K_s} n_{ij}(t), \quad 1 \leq s \leq r. \tag{1.43}$$

Theorem 1.4. *The stationary distribution of the process $z'_2(t)$ is calculated by the formula*

$$P'_2(n_i, 1 \leq i \leq m, N_s, 1 \leq s \leq r) =$$

$$= C_2^{-1} P'(n_1, ..., n_m) \prod_{s=1}^{r} \Pi_s(N_s), \tag{1.44}$$

$$n_i \geq 0, \ N_s \geq 0, \ 1 \leq i \leq m, \ 1 \leq s \leq r, \ \sum_{i=1}^{m} n_i + \sum_{s=1}^{r} N_s = M,$$

$$C_2 = \sum_{n_i \geq 0,\, N_s \geq 0,\, \sum_{i=1}^{m} n_i + \sum_{s=1}^{r} N_s = M.} P'(n_1, ..., n_m) \prod_{s=1}^{r} \Pi_s(N_s),$$

$\Pi_s(N_s)$ *are calculated by the formulas (1.40).*

Proof. This theorem proof repeats the theorem 1.3 proof.

1.4. Product theorems for queuing networks with prohibitions

In this subsection the product theorems obtain new direction of a development. This direction is connected with the product form stationary distributions of queueing networks with prohibitions. Last years this field of an investigation attracts special attention [35], [46], [50], [60].

Opened and closed queueing networks are considered in terms of graphs, whose nodes are states of Markov processes, described these networks. Edges of these graphs correspond to nonzero transition intensities of the Markov processes. A special choice of motion equations (as traditional so modified ones) allow to obtain the product forms of the stationary distributions for the considered Markov processes. An algorithm of route matrices construction in the networks with modified motion equations is suggested.

Opened networks

Consider opened Jackson network G with m nodes (oneserver queueing systems of the $M|M|1|\infty$ type), the input flow intensity $\lambda > 0$, the serving intensities $\mu_1 > 0, \ldots, \mu_m > 0$ and the route matrix $\Theta = (\theta_{ij})_{i,j=0}^{m}$:

$$\theta_{00} = 0, \ \theta_{ki} > 0, \ 0 \le k, i \le m. \tag{1.45}$$

Here θ_{ki} is the probability of the transition from the k-th node to the i-th node (the node with the index 0 is external source which creates input flow).

Denote the base vectors in the linear space E^m by $\mathbf{e}_1 = (1, 0, \ldots, 0)$, $\mathbf{e}_2 = (0, 1, 0, \ldots, 0), \ldots, \mathbf{e}_m = (0, \ldots, 0, 1)$ and designate the set $Z^m = \mathbf{N} = (n_1, \ldots, n_m)$, $n_1 \ge 0, \ldots, n_m \ge 0$. The vector of the customers current numbers in the network G nodes is described by the Markov process $N(t)$, $t \ge 0$ with the state set $\mathcal{L} = Z^m$ and with nonzero transition intensities

$$L(\mathbf{N}, \mathbf{N} + \mathbf{e}_k) = \lambda \theta_{0k}, \ \ \mathbf{N} \in Z^m, \ \ L(\mathbf{N}, \mathbf{N} - \mathbf{e}_k) = \mu_k \theta_{k0}, \tag{1.46}$$

$$L(\mathbf{N}, \mathbf{N} - \mathbf{e}_k + \mathbf{e}_i) = \mu_k \theta_{ki}, \ \ 1 \le k \ne i \le m, \ \mathbf{N} \in Z^m, \ n_k > 0. \tag{1.47}$$

As the route matrix Θ is indivisible:

$$\forall i, j \in \{0, 1, ..., m\} \exists i_1, ..., i_r \in \{1, ..., m\} : \ \theta_{ii_1} > 0, \ \theta_{i_1 i_2} > 0, \ ..., \theta_{i_r j} > 0,$$

so the system of the motion equations

$$(\lambda_1, \ldots, \lambda_m) = (\lambda, \lambda_1, \ldots, \lambda_m)\Theta_0 \tag{1.48}$$

has the unique solution and $\lambda_1 \geq 0, \ldots, \lambda_m \geq 0$. Here the matrix Θ_0 is obtained from the matrix Θ by the deleting of 0-indexed column. If $\lambda_1 < \mu_1, \ldots, \lambda_m < \mu_m$ then the discrete Markov process $N(t)$, $t \geq 0$, is ergodic [34] and its stationary distribution [36] is calculated by the formula

$$P(\mathbf{N}) = C^{-1}\Phi(\mathbf{N}), \quad \Phi(\mathbf{N}) = \prod_{i=1}^{m}\left(\frac{\lambda_i}{\mu_i}\right)^{n_i}, \tag{1.49}$$

$$C = \sum_{\mathbf{N}\in\mathcal{L}} \Phi(\mathbf{N}), \quad \mathbf{N} \in \mathcal{L}.$$

To describe the network G with the prohibitions consider non-directed graph Γ with the nodes from state set $\mathcal{L} \subseteq Z^m$ and with the edges represented by the formulas

$$[\mathbf{N}, \mathbf{N} + \mathbf{e}_k], \quad \mathbf{N} \in Z^m, \quad 1 \leq k \leq m, \tag{1.50}$$

$$[\mathbf{N}, \mathbf{N} - \mathbf{e}_k + \mathbf{e}_i], \quad \mathbf{N} \in Z^m, \quad n_k > 0, \quad 1 \leq k \neq i \leq m. \tag{1.51}$$

If the edge of the (1.50) type belongs to the graph Γ then an input customer (from the node 0) may arrive to the $k-$th node and a customer from the $k-$th node may depart from the network (may arrive in the $0-$th node). If the edge of the type (1.51) belongs to the graph Γ then the customer may depart from the $i-$th node and arrive in the $k-$th node and vice-versa. An absence of the the types (1.50), (1.51) edges in the graph Γ means that the corresponding transitions (in the both sides) are prohibited.

If the edge of the type (1.50) is absent then an input customer, arriving in the $k-$th node, departs from the network and the customer, departing from the $k-$th node, returns in this node. If the edge of the type (1.51) is absent then the customer, moving from the $i-$th edge to the $k-$th edge, returns to the $i-$th edge and vice-versa the customer, moving from the $k-$th edge to the $i-$th edge, returns to the $k-$th edge. The prohibition of the type (1.50) edge (of of the type (1.51) edge) means that the transition intensity (1.46) (the transition intensity (1.47)) becomes equal to zero. These allowances and prohibitions define a protocol of the network

$G.$ Such protocols permit in the network G a return of rejected customers and so an origin of retrial queues.

Theorem 1.5. *Suppose that $\mathcal{L}_0$ is finite set in Z^m so that the graph Γ, consisted of the edges $\{[\mathbf{N}, \mathbf{N} + \mathbf{e}_i],\ [\mathbf{N} + \mathbf{e}_i, \mathbf{N} + \mathbf{e}_k] :\ \mathbf{N} \in \mathcal{L}_0,\ 1 \leq i \neq k \leq m\}.$ is connected. Then the discrete Markov process $N(t)$ with the state set $\mathcal{L} = \mathcal{L}_0 \bigcup \{\mathbf{N} + \mathbf{e}_i :\ \mathbf{N} \in \mathcal{L}_0,\ 1 \leq i \leq m\}$, describing the network G with the prohibitions, is ergodic and its limit distribution $P(\mathbf{N})$, $\mathbf{N} \in \mathcal{L}$, is calculated by the formula (1.49).*

Proof. Denote

$$F_0(\mathbf{N}) = \{\mathbf{N} + \mathbf{e}_i,\ 1 \leq i \leq m\},$$

$$F_i(\mathbf{N}) = \{\mathbf{N} - \mathbf{e}_i,\ \mathbf{N} - \mathbf{e}_i + \mathbf{e}_j,\ 1 \leq j \leq m,\ j \neq i\},$$

$$1 \leq i \leq m,\ \mathbf{N} \in Z^m.$$

It is clear that

$$F_i(\mathbf{N}) \bigcap F_j(\mathbf{N}) = \emptyset. \tag{1.52}$$

If the conditions of the theorem 1.5 are true then in the graph Γ for each $\mathbf{N} \in \mathcal{L}$ there exists the index set $I(\mathbf{N}) \subseteq \{0, \ldots, m\}$ so that the set $S(\mathbf{N})$ of the nodes, connected with the node $\mathbf{N}$ by one edge paths, is represented as follows

$$S(\mathbf{N}) = \bigcup_{i \in I(\mathbf{N})} F_i(\mathbf{N}). \tag{1.53}$$

From the formulas (1.46), (1.47), (1.48) obtain

$$\sum_{\mathbf{J} \in F_i(\mathbf{N})} [\Phi(\mathbf{N})L(\mathbf{N}, \mathbf{J}) - L(\mathbf{J}, \mathbf{N})\Phi(\mathbf{J})] = 0, \tag{1.54}$$

$$0 \leq i \leq m,\ \mathbf{N} \in \mathcal{L}.$$

Using the equalities (1.54) obtain for $\mathbf{N} \in \mathcal{L}$:

$$\sum_{\mathbf{J} \in \mathcal{L}} [\Phi(\mathbf{N})L(\mathbf{N}, \mathbf{J}) - L(\mathbf{J}, \mathbf{N})\Phi(\mathbf{J})] =$$

$$= \sum_{\mathbf{J} \in S(\mathbf{N})} [\Phi(\mathbf{N})L(\mathbf{N}, \mathbf{J}) - L(\mathbf{J}, \mathbf{N})\Phi(\mathbf{J})] = 0. \tag{1.55}$$

As the graph Γ is connected and the formulas (1.45) (1.55) are true and transition intensities $L(\mathbf{N}, \mathbf{J})$ are bounded by the quantity $\lambda + \mu_1 + \ldots + \mu_m$ then the process $N(t)$. describing the opened queueing network with the state set $\mathcal{L}$ and the graph Γ, satisfies the conditions of the well known ergodicity theorem [34]. So the process $N(t)$ is ergodic and its limit distribution $P(\mathbf{N})$ satisfies the formula (1.49).

As examples of the set $\mathcal{L}_0$, satisfying the conditions of the theorem 1.5, we may take the following sets: $\{\mathbf{N} \in Z^m : M' \leq \sum_{k=1}^{m} n_k < M\}$, $\{\mathbf{N} \in Z^m : M'_k \leq n_k < M_k, \ 1 \leq k \leq m\}$, $0 \leq M', M, M'_1, M_1, \ldots,$ $M'_m, M_m < \infty$.

Theorem 1.6. *Suppose that $\mathcal{L}$ is the finite set in Z^m, Γ is the connected graph with the node set $\mathcal{L}$ and with the types (1.50), (1.51) edges. If the matrix Θ elements and the numbers $\lambda_1 > 0, \ldots, \lambda_m > 0$ satisfy the equalities*

$$\lambda_k \theta_{k0} = \lambda \theta_{0k}, \quad 1 \leq k \leq m, \tag{1.56}$$

$$\lambda_i \theta_{ik} = \lambda_k \theta_{ki}, \quad 1 \leq k \neq i \leq m, \tag{1.57}$$

then the discrete Markov process $N(t)$ with the state set $\mathcal{L}$, describing the network G with the prohibitions, is ergodic and its limit distribution $P(\mathbf{N}), \mathbf{N} \in \mathcal{L}$, is calculated by the formula (1.49).

Proof. The equalities (1.49), (1.56), (1.57) lead to

$$\Phi(\mathbf{N})\lambda \theta_{0k} - \Phi(\mathbf{N} + \mathbf{e_k})\mu_k \theta_{k0} = 0, \tag{1.58}$$

$$\mathbf{N}, \ \mathbf{N} + \mathbf{e_k} \in \mathcal{L}, \ 1 \leq k \leq m,$$

$$\Phi(\mathbf{N})\mu_i \theta_{ik} - \Phi(\mathbf{N} + \mathbf{e_k} - \mathbf{e_i})\mu_k \theta_{ki} = 0, \tag{1.59}$$

$$\mathbf{N}, \mathbf{N} + \mathbf{e_k} - \mathbf{e_i} \in \mathcal{L}, 1 < k \neq i < m.$$

As the formulas (1.46), (1.47), (1.58), (1.59) are true then the function $\Phi(\mathbf{N})$ satisfies the equations (1.55). The end of theorem 1.6 proof practically coincides with the end of theorem 1.5 proof.

Closed networks

Consider now closed Jackson network G' with m nodes (the oneserver queueing systems of the type $M|M|1|\infty$ with the serving intensities $\mu_i > 0$, $i = 1\ldots,m$), the route matrix $\Theta' = (\theta'_{ki})_{k,i=1}^{m}$ so that

$$\theta'_{ki} > 0, \ 0 \le k, i \le m. \tag{1.60}$$

The network G' is described by the discrete Markov process $N(t)$, $t \ge 0$, with the state set $Z = \{\mathbf{N} \in Z^m, \sum_{i=1}^{m} n_i = M\}$ (M is the number of network customers) and with the nonzero transition probabilities (1.47).

As the route matrix $\Theta' = (\theta'_{ki})_{k,i=1}^{m}$ of the network G' satisfies the condition (1.60) so for each $B > 0$ the appropriate system of the motion equations [33]:

$$(\lambda_1, \ldots, \lambda_m) = (\lambda_1, \ldots, \lambda_m)\Theta', \ B = \sum_{i=1}^{m} \lambda_i \tag{1.61}$$

has the unique solution $\lambda_1 \ge 0, \ldots, \lambda_m \ge 0$. Then the process $N(t)$ is ergodic [34] and its limit distribution [33] is calculated by the formula (1.49).

Analogously correspond to the network G' the graph Γ with the node set $\mathcal{L} \subseteq Z$ and the edges of the type (1.51). Formulate the product theorems for the network G' with the prohibitions.

Theorem 1.7. *Suppose that $\mathcal{L}_0$ is finite set in Z_m so that the graph Γ, consisted of the edges $\{[\mathbf{N} + \mathbf{e}_i, \mathbf{N} + \mathbf{e}_k] : \ \mathbf{N} \in \mathcal{L}_0, \ 1 \le i \ne k \le m\}$, is connected. Then the discrete Markov process $N(t)$ with the state set $\mathcal{L} = \{\mathbf{N} + \mathbf{e}_i : \mathbf{N} \in \mathcal{L}_0, 1 \le i \le m\}$, describing the network G' with the prohibitions, is ergodic and its limit distribution $P(\mathbf{N})$, $\mathbf{N} \in \mathcal{L}$, is calculated by the formula (1.49).*

Proof. The theorem 1.7 proof practically the word by the word repeats the theorem 1.5 proof. Changes are following: the inclusion $I(\mathbf{N}) \subseteq \{0, \ldots, m\}$ is replaced by the inclusion $I(\mathbf{N}) \subseteq \{1, \ldots, m\}$ and in the the formula (1.54) the inequality $0 \le i \le m$ is replaced by the inequality $1 \le i \le m$.

Theorem 1.8. *Suppose that $\mathcal{L}$ is the finite set in Z_m, Γ is the connected graph with the node set $\mathcal{L}$ and with the edges (1.51). If the matrix Θ' elements and the numbers $\lambda_1 > 0, \ldots, \lambda_m > 0$ satisfy the equalities*

$$\lambda_i \theta'_{ik} = \lambda_k \theta'_{ki}, \ 1 \le k, i \le m, \tag{1.62}$$

then the discrete Markov process $N(t)$ with the state set $\mathcal{L}$, describing the network G' with the prohibitions, is ergodic and its limit distribution $P(\mathbf{N})$, $\mathbf{N} \in \mathcal{L}$, is calculated by the formula (1.49).

Proof. The equalities (1.49), (1.62) lead to the formula (1.59). So the function $\Phi(\mathbf{N})$ satisfies the equations (1.55). The end of the theorem 1.8 proof repeats the end of the theorem 1.5 proof.

Remark 6. *If in the conditions of the theorems 1.5 - 1.8 $\lambda_1 < \mu_1, ...,$ $\lambda_m < \mu_m$ then the set $\mathcal{L}_0$ may be infinite.*

Remark 7. *New results in the theorem 1.6 (in the theorem 1.8) are obtained not by special choice of the graph Γ (an arbitrary connected graph). These results are obtained with the replacement of the Jackson motion equations (1.48) (of the Gordon-Newel motion equations (1.61)) by the motion equations (1.56), (1.57) (the motion equations (1.62)).*

Remark 8. *The proof of the theorem 1.5 leads to the statement that this theorem is true for any connected graph Γ, which satisfies the following condition. For each node $\mathbf{N}$ of the graph Γ there exists the index set $I(\mathbf{N}) \subseteq \{0, \ldots, m\}$ so that the set $S(\mathbf{N})$ of the nodes, connected with the node $\mathbf{N}$ by one edge paths, is represented in the form (1.53). It is not difficult to prove that the theorem 1.5 conditions describe all graphs Γ, satisfying this condition. Analogous statement is true for the theorem 1.7 too.*

Algorithm of route matrix construction

Opened network. Suppose that $\lambda > 0$, $\theta_{0k} > 0$, $1 > \theta_{k0} > 0$, $1 \leq k \leq m$, are fixed. Using the equalities (1.56) define $\lambda_1, \ldots, \lambda_m$ and θ_{ki}, $1 \leq k, i \leq m$, which are the following problem solutions.

Denote

$$\Lambda_k = \lambda_k(1 - \theta_{k0}), \tag{1.63}$$

then from (1.56)

$$\Lambda_k = \frac{\lambda \theta_{0k}(1 - \theta_{k0})}{\theta_{k0}}, \quad 1 \leq k \leq m. \tag{1.64}$$

Take

$$\pi_{ki} = \theta_{ki}/(1 - \theta_{k0}), \tag{1.65}$$

consequently

$$\sum_{i=1}^{m} \pi_{ki} = 1, \quad 1 \le k \le m. \tag{1.66}$$

Denote

$$C_{ki} = \Lambda_k \pi_{ki}, \quad 1 \le k, i \le m. \tag{1.67}$$

As the formulas (1.57), (1.66) are true so the matrix $C = (C_{ki})_{k,i=1}^{m}$ is the permissible solution of transportation problem

$$\sum_{i=1}^{m} C_{ki} = \Lambda_k, \quad C_{ki} = C_{ik} > 0, \quad 1 \le k, i \le m. \tag{1.68}$$

So using fixed $\lambda > 0, \theta_{0k} > 0, 1 > \theta_{k0} > 0, 1 \le k \le m$, find from the formulas (1.63) $\Lambda_k, 1 \le k \le m$. Then using $\Lambda_k, 1 \le k \le m$, define the permissible solution of the transportation problem (1.68). The formulas (1.65), (1.67) allow to calculate $\pi_{ki}, \theta_{ki}, 1 \le k, i \le m$.

Consider now the permissible solutions of the transportation problem (1.68). Renumber the nodes $1, \ldots, m$ so that $\Lambda_1 \le \Lambda_2, \ldots, \Lambda_1 \le \Lambda_m$. Choose $C_{11}, \ldots, C_{1m}$ from the conditions

$$\sum_{k=1}^{m} C_{1k} = \Lambda_1, \quad C_{11} > 0, \ldots, C_{1m} > 0$$

and take

$$C_{k1} = C_{1k}, \quad k = 1, \ldots, m.$$

Redefine now $\Lambda_2, \ldots, \Lambda_m$ by

$$\Lambda_2 := \Lambda_2 - C_{21}, \ldots, \Lambda_m := \Lambda_m - C_{m1}.$$

So it is necessary only to define $C_{ki}, 2 \le k, i \le m$, satisfying the formula

$$\sum_{i=2}^{m} C_{ki} = \Lambda_k, \quad C_{ki} = C_{ik} > 0, \quad 2 \le k, i \le m. \tag{1.69}$$

Apply this procedure but to the problem (1.69) and so on.

So the constructed algorithm is the modified version of the well known North-West method [27]. This method allows to find the permissible solutions of the transportation problem with the additional condition (1.68) that the matrix C is symmetric.

Closed network. Suppose that the intensities $\lambda_1 > 0, \ldots, \lambda_m > 0$ are fixed. Denote

$$C'_{ki} = \lambda_k \theta'_{ki}, \quad 1 \leq k, i \leq m, \tag{1.70}$$

from (1.62) and the equalities

$$\sum_{i=1}^{m} \theta'_{ki} = 1, \quad k = 1, \ldots, m,$$

obtain that the matrix $C' = (C'_{ki})_{k,i=1}^{m}$ is the permissible solution of the transportation problem

$$\sum_{i=1}^{m} C'_{ki} = \lambda_k, \quad C'_{ki} = C'_{ik} > 0, \quad 1 \leq k, i \leq m. \tag{1.71}$$

So to find the matrix C' use the previous algorithm. Then, using the known matrix C' and the fixed intensities $\lambda_1, \ldots, \lambda_m$, find from (1.70) the route matrix Θ'.

Networks with random varying structure.

Suppose that the opened queueing system G has the graph Γ, satisfying the theorem 1.5 conditions, and randomly varying from one state to another state. In the i-th state the network G has the graph Γ_i with the state set $\mathcal{L}$ and the edge set $\mathcal{L}_0 = \mathcal{L}_0^i$. The network in the i-th state is described by the Markov process $N_i(t)$ with the state set $\mathcal{L} = \mathcal{L}_0^i \bigcup \{\mathbf{N} + \mathbf{e}_i : \mathbf{N} \in \mathcal{L}_0^i, 1 \leq i \leq m\}$ and the transition intensities $L_i(\mathbf{N}, \mathbf{N}^*), \mathbf{N}, \mathbf{N}^* \in \mathcal{L}$. Its limit distribution $P(\mathbf{N})$ is calculated by the formula (1.49).

If the constructed queueing network with the randomly varying structure may be described by the discrete Markov process $z(t)$ with the state set $\mathcal{L} \times I$, $I = \{1, 2, ..., K\}$, and the transition intensities

$$\mathbf{L}((\mathbf{N}, i), (\mathbf{N}, j)) = \Upsilon_{\mathbf{N}}(i, j),$$

$$\mathbf{L}((\mathbf{N}, i), (\mathbf{N}^*, i)) = L_i(\mathbf{N}, \mathbf{N}^*), \quad \mathbf{N}, \mathbf{N}^* \in \mathcal{L}, \ i, j \in I$$

then the following theorem is true.

Theorem 1.9. *If for fixed* $\mathbf{N}$ *the elements of the matrix* $\|\Upsilon_{\mathbf{N}}(i,j)\|_{i,j\in I}$ *are positive solutions of the system*

$$A_i \sum_{j\in I} \Upsilon_{\mathbf{N}}(i,j) = \sum_{j\in I} A_j \Upsilon_{\mathbf{N}}(j,i), \ i \in I,$$

where $A_i > 0$, $i \in I$, $\sum_{i\in I} A_i = 1$, *then the process* $z(t)$ *is ergodic and its limit distribution is calculated by the formula*

$$\mathbf{P}(\mathbf{N},i) = A_i P(\mathbf{N}).$$

Proof. This statement is the direct consequence of the theorem 2 from [56].

The figure 5 contains possible graphs for different states of the network G.

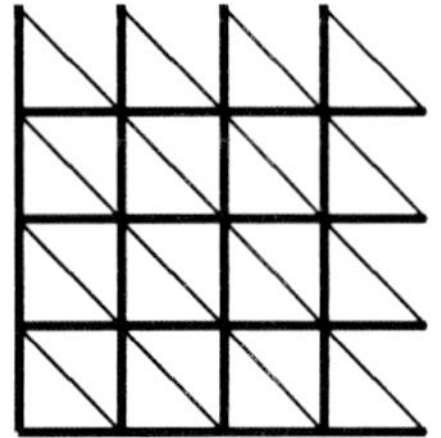 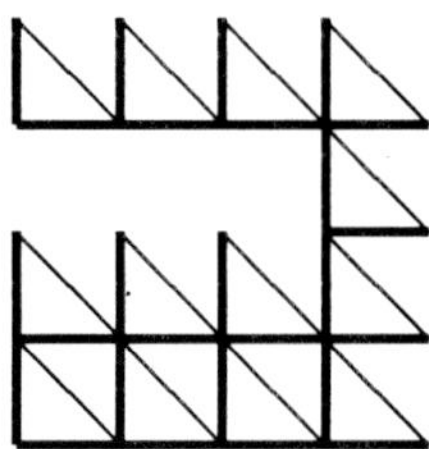

Fig. 5

Remark 9. *Analogous theorem may be proved for the networks in which the graphs satisfy the theorem 1.6 conditions. In this case the graphs* $\Gamma_1, ..., \Gamma_K$ *differ from each other only by edges. The quantities* λ_i *in limit distribution* $P(\mathbf{N})$, *calculated by the formula (1.49), satisfy the formulas (1.56), (1.57).*

1.5.   Stochastic control of Markov process parameter

In the previous subsections the effects of the counteraction between independent subsystems of a queueing system are considered. In this subsection queueing systems with conditionally independent subsystems are considered.

For this aim the discrete Markov processes with randomly varying parameters are considered. Stochastic control of their parameters are constructed. Then the resulted stationary distributions are probability mixtures of the processes with fixed meanings of the parameter distributions. These problem is new in a development of the product theorems.

Main result of this subsection is a construction of one to one transformation from a set of all possible stochastic controls of Markov process parameter and a set of permissible solutions of some transportation problem.

Connection between stochastic control of discrete Markov process parameter and transportation problem

Consider the Markov processes $x^{(i)}(t)$, $t \geq 0$, $i \in I = \{1, ..., N\}$, with the matrices of the transition intensities $||\mu_{t,j}^{(i)}||_{t,j \in J}$, J is the finite set, consisted of nonnegative elements out of its diagonal and zeros at the diagonal. Suppose that the matrices of the transition intensities satisfy the following condition:

$(\mathbf{B}^*)$ for $\forall i \in I$

$$\forall j, t \in J \; \exists j_1, ..., j_r \in J : \; \mu_{j,j_1}^{(i)} > 0, \; \mu_{j_1,j_2}^{(i)} > 0, ..., \mu_{j_r,t}^{(i)} > 0.$$

Then [34] the processes $x^{(i)}(t)$, $t \geq 0$, $i \in I$, are ergodic and possess the limit distributions $\overline{B}_j^{(i)}$, $j \in J$, $i \in I$:

$$\overline{B}_j^{(i)} \sum_{t \in J} \mu_{j,t}^{(i)} = \sum_{t \in J} \overline{B}_t^{(i)} \mu_{t,j}^{(i)}, \quad \sum_{j \in J} \overline{B}_j^{(i)} = 1. \tag{1.72}$$

Consider the Markov process $z(t)$, $t \geq 0$, with nonzero transition intensities

$$\gamma((i,j),(i,t)) = \mu_{j,t}^{(i)}, \; \gamma((i,j),(k,j)) = \lambda_{i,k}^{(j)}, \; j, t \in J, \; k, i \in I, \tag{1.73}$$

satisfied the following condition: the matrices $||\lambda_{i,k}^{(j)}||_{i,k \in I}$, $j \in J$, consist of zero diagonal elements and positive elements out of the diagonal (see the condition $(\mathbf{C}^*)$). Then the Markov process $z(t)$, $t \geq 0$, satisfies the condition $(\mathbf{B}^*)$ and so is ergodic. Denote by $\overline{\Psi}_{i,j}$, $i \in I$, $j \in J$, the limit distribution of this process.

Find the control $\overline{U} = (\Lambda^{(j)} = ||\lambda_{k,i}^{(j)}||_{k,i \in I}$, $j \in J)$, which consists of the matrices, satisfying the condition $(\mathbf{C}^*)$ and the equalities

$$\overline{\Psi}_{i,j} = A_i \overline{B}_j^{(i)}, \; i \in I, \; j \in J. \tag{1.74}$$

Denote $\overline{F}_k^{(j)} = A_k \overline{B}_j^{(k)} > 0$ and put

$$\overline{G}_{k,i}^{(j)} = \overline{F}_k^{(j)} \lambda_{k,i}^{(j)}, \; i \in I, \; k \in I, \; j \in J. \tag{1.75}$$

Theorem 1.10. *The formula (1.75) defines one to one transformation of the set of the controls $\overline{U}$, which satisfy the condition* $(\mathbf{C^*})$ *and the equalities (1.74), on the set of all sequences* $\overline{V} = (\|\overline{G}_{k,\,i}^{(j)}\|_{k,\,i\in I},\ j \in J)$ *of the matrices, which satisfy the condition* $(\mathbf{C^*})$ *and are the permissible solutions of transportation problems:*

$$\sum_{k\in I}\overline{G}_{i,\,k}^{(j)}=D_i^{(j)},\ \ i\in I, \tag{1.76}$$

$$\sum_{k\in I}\overline{G}_{k,\,i}^{(j)}=D_i^{(j)},\ \ i\in I,\ \ j\in J. \tag{1.77}$$

for some $D_i^{(j)} > 0,\ i\in I,\ j\in J.$

Proof. Suppose that the matrices from the control $\overline{U}$ satisfy the condition (1.74). Define $D_i^{(j)} > 0,\ i\in I,\ j\in J$, by the formula (1.76) and obtain the formula (1.77). For this aim, using the formula (1.74), obtain

$$\sum_{k\in I,\,t\in J} A_i\overline{B}_j^{(i)}\gamma((i,j),(k,t))-\sum_{k\in I,\,t\in J} A_k\overline{B}_t^{(k)}\gamma((k,t),(i,j))=0. \tag{1.78}$$

Put the equalities (1.73) into (1.78) and obtain

$$\sum_{k\in I} A_i\overline{B}_j^{(i)}\lambda_{i,k}^{(j)} + \sum_{t\in J} A_i\overline{B}_j^{(i)}\mu_{j,\,t}^{(i)}-$$

$$-\sum_{k\in I} A_k\overline{B}_j^{(k)}\lambda_{k,i}^{(j)} - \sum_{t\in J} A_i\overline{B}_t^{(i)}\mu_{t,\,j}^{(i)} = 0. \tag{1.79}$$

Using the formula (1.72), obtain from (1.79) the formula (1.77).

Suppose now that for some $D_i^{(j)}>0,\ i\in I, j\in J$, the matrices $\|\overline{G}_{k,\,i}^{(j)}\|_{k,\,i\in I}, j\in J$, satisfy the condition $(\mathbf{C^*})$ and the formulas (1.76), (1.77). Using the formula (1.75) define the control $\overline{U} = \Lambda^{(j)} = (\|\lambda_{k,\,i}^{(j)}\|_{k,\,i\in I},\ j\in J)$. As the matrices $\Lambda^{(j)},\ j\in J$, satisfy the condition $(\mathbf{C^*})$ then the states of the Markov process $z(t),\ t\geq 0$, are connected in the condition $(\mathbf{B^*})$ sense. The formulas (1.76), (1.77) lead to the equalities

$$\sum_{k\in I} A_i\overline{B}_j^{(i)}\lambda_{i,k}^{(j)} - \sum_{k\in I} A_k\overline{B}_j^{(k)}\lambda_{k,i}^{(j)} = 0. \tag{1.80}$$

The formulas (1.72), (1.80) lead to (1.79) and so to the formula (1.78). As the result the process $z(t),\ t\geq 0$, is ergodic with the limit distribution (1.74).

Stochastic control of the system $M|M|1|n$ parameter

As an example consider the system $M|M|1|n$ with single server, n places in a queue, Poisson input flow with the intensity $a > 0$ and serving time with the exponential distribution and the parameter $b > 0$. Then the stationary distribution of the customers number $\overline{B}_j$, $j \in J = \{0, ..., n+1\}$, in the system is following:

$$\overline{B}_j = \frac{\rho^j(1-\rho)}{(1-\rho^{n+1})}, \quad 0 \le j \le n+1, \ \rho = a/b.$$

Suppose that there are two different vectors $(a^{(i)}, b^{(i)})$, $i \in I = \{1, 2\}$, with positive components and define the stationary distribution of customers number in the system, corresponding these vectors:

$$\overline{B}_j^{(i)} = \frac{\rho_i^j(1-\rho_i)}{(1-\rho_i^{n+1})}, \quad 0 \le j \le n+1, \ \rho_i = \frac{a_i}{b_i}, \ i = 1, 2.$$

Using the vectors $(a^{(i)}, b^{(i)})$, $i \in I$, it is possible to define the transition intensities matrices $||\mu_{t,j}^{(i)}||_{t,j \in J}$, $i \in I$, of the processes $x^{(i)}(t)$, $t \ge 0$, $i \in I$, corresponding these vectors and defining customers numbers in the system $M|M|1|n$ with these parameters.

Fix positive numbers D, A_1, A_2 so that $A_1 + A_2 = 1$. Define $\lambda_{1,2}^{(j)}$, $\lambda_{2,1}^{(j)}$ from the equalities

$$D = A_1 \overline{B}_j^{(1)} \lambda_{1,2}^{(j)} = A_2 \overline{B}_j^{(2)} \lambda_{2,1}^{(j)}, \quad j = 0, ..., n+1.$$

As a result obtain the matrices $||\lambda_{i,k}^{(j)}||_{i,k \in I}$, $j \in J$, which satisfy the conditions (1.76), (1.77) with $D_i^{(j)} = D$, $i \in I$, $j \in J$. Define the process $z(t)$, $t \ge 0$, describing the system $M|M|1|n$ with input intensities, which change accordingly to the transition intensities matrices $||\lambda_{i,k}^{(j)}||_{i,k \in I}$, $j \in J$. The stationary distribution $\overline{\Psi}_{i,j}$ of the process $z(t)$ satisfy the formula (1.74).

Algorithm of stochastic control matrix for Markov process

Consider now how to generate arbitrary matrices, which are the permissible solutions of the transportation problem occurred in the theorem 1.10. We construct the set of all matrix sequences $\overline{V} = (||G_{k,i}^{(j)}||_{k,i \in I}, j \in J)$, which consist of zero diagonal and positive out of the diagonal elements and satisfy the conditions

$$\sum_{k \in I} \overline{G}_{i,k}^{(j)} = \sum_{k \in I} \overline{G}_{k,i}^{(j)}, \quad i \in I, \ j \in J. \tag{1.81}$$

Fix $j \in J$ and denote $\|\overline{G}^{(j)}_{k,\,i}\|_{k,\,i\in I} = \|g_{k,\,i}\|_{k,\,i\in I} = \mathbf{g}$ and represent $\mathbf{g}$ in the following form: $\mathbf{g} = \mathbf{g^s} + \mathbf{g^a}$, where $\mathbf{g^s}$ is symmetric and $\mathbf{g^a}$ is antisymmetric components of the matrix $\mathbf{g}$. The construction of arbitrary matrix $\mathbf{g}$, satisfying the formula (1.81), algorithm consists of the following steps. Take arbitrary symmetric matrix $\mathbf{g^s} = \|g^s_{i,\,k}\|$ with positive elements out of the diagonal and zeros at the diagonal. Then choose arbitrary antisymmetric matrix $\mathbf{f^a} = \|f_{i,\,k}\|$ so that

$$\sum_{k\in I} f_{i,\,k} = 0, \quad f_{i,\,k} \neq 0, \quad i \neq k. \tag{1.82}$$

To construct the matrix $\mathbf{f^a}$ choose arbitrary real numbers $f_{i,\,k}$, $i < k < m = |I|$ (here $|I|$ is the number of the elements in the finite set I) and then calculate $f_{i,\,m}$ from (1.82). Define

$$F = \min_{i\neq k} \frac{g^s_{i,\,k}}{|f_{i,\,k}|}$$

and for an arbitrary number R, $|R| < F$, take

$$g^a_{i,\,k} = Rf_{i,\,k}, \quad g_{i,\,k} = g^s_{i,\,k} + g^a_{i,\,k}.$$

It is trivial to prove that this algorithm gives general solution of the problem (1.81).

Chapter 2

Cooperative effects in risk models with discrete time

Known mathematical models of insurance [20, 39] are characterized by the following parameters: the initial capital x, the ruin probability $p(x)$ and the insurance percent b and the distribution of the risk (the damage, the loss). Usually a behavior of the function $p(x)$ is investigated in the case of a fixed but not small b and a large x. Modern applied insurance systems of natural catastrophes: floods, droughts, forest fires, earthquakes, tsunami and etc., demand introducing changes to such a formulation of the problem. A necessity of a construction and an investigation of insurance systems with a small insurance percent b, a small initial capital x and a small ruin probability $p(x)$ appears in the case of large risks. Existing risk models [20, point 8.7] do not possess these properties.

In this chapter, a risk model satisfying these properties is constructed. It is based on the principle of a mutual insurance that is a considered system and is an aggregation of n independent and identical insurance systems. It is possible not only to recognize the cooperative effects in the aggregated system but to extract in the parameter set the regions, where for $n \to \infty$ these effects are significant, and the regions, where the effects are small. The specifics of suggested model of mutual insurance with independent and weak dependent risks is the existence of clear boundary between these regions.

The Mutual insurance model is closely connected with oneserver queueing system

$GI|GI|1|\infty$ in a regime of a high load. A ratio between a "size" of a small fluctuation and a deviation of the queueing system load coefficient from the limit meaning 1 lead to the effect of sharp transition from one regime to another regime.

In comparison with the previous chapter, where the calculations and the proves are based on the combinatorial formulas, the motion equations and the algorithms of the transportation problem solutions, the results of this chapter are based on the probability theory limit theorems and on the estimates of the rate convergence in these theorems.

2.1. Model of mutual insurance with independent risks

Consider n independent and identical insurance companies. Suppose that the annual risk of the j-th company in the k-th year is $x(k, j)$ and the random variables (r.v.'s) $x(k, j)$, $j = 1, \ldots, n$, $k = 1, 2, \ldots$, are independent and identically distributed,

$$M\, x(k, j) = 1, \ p(x(k, j) < t) = G(t).$$

Suppose that the annual prizes of the single company equal to $1+b$, where $b = n^{-\gamma}$, $\gamma > 0$. As the common prize of n companies aggregation is $(1 + b)n$ and the common risk is $\sum_{j=1}^{n} x(k, j)$ so the ruin probability

$$p_n = p_n(x) = P\left(\sup_{m>0} \sum_{k=1}^{m} \sum_{j=1}^{n} (x(k, j) - 1 - b) > x \right). \tag{2.1}$$

Suppose that x is a fixed and sufficiently small quantity, for example $x = 0$.

Consider a case of large risks. The risk is large if [52] its distribution function (d.f.) $G(t) = 0$, $t \leq 0$, for some α, C, $1 < \alpha < 2, C > 0$, satisfies the condition

$$1 - G(t) \sim \frac{C(2 - \alpha)}{\alpha}\, t^{-\alpha}, \quad t \to +\infty. \tag{2.2}$$

Theorem 2.1. *Suppose that for some α, $1 < \alpha < 2$, there exists $C > 0$ so that d.f. G satisfies the condition (2.2). If the inequality*

$$\gamma < 1 - \frac{1}{\alpha} \tag{2.3}$$

is true then for any τ, $1/(1-\gamma) < \tau < \alpha$, there exists the positive number $C_1 = C_1(\tau)$ so that

$$p_n \leq C_1 n^{1 - \tau(1-\gamma)}, \quad n = 1, 2, \ldots \tag{2.4}$$

If the inequality (2.3) is not true then

$$\lim_{n \to \infty} \inf(p_n, \ n \geq 1) > 0. \tag{2.5}$$

Corollary 2.1. *If the inequality (2.3) is true then*

$$\lim_{n \to \infty} p_n = 0. \tag{2.6}$$

Theorem 2.2. *Suppose that d.f. G satisfies the theorem 2.1 conditions. If the inequality*

$$\gamma > 1 - \frac{1}{\alpha} \tag{2.7}$$

is true then

$$\lim_{n \to \infty} p_n = 1. \tag{2.8}$$

For a comparison consider a case when the risks are not large (d.f. G has a finite variation and does not satisfies the condition (2.2)):

$$D\,x(k,j) = \sigma^2, \quad 0 < \sigma^2 < \infty. \tag{2.9}$$

Theorem 2.3. *Suppose that d.f. G satisfies the condition (2.9). If the inequality*

$$\gamma < \frac{1}{2} \tag{2.10}$$

is true then there exists the positive number C_2 so that

$$p_n \leq C_2 n^{2\gamma - 1}, \quad n = 1, 2, \ldots \tag{2.11}$$

If the condition (2.10) is not true then the formula (2.5) takes place.

Corollary 2.2. *If the inequality (2.10) is true then the formula (2.6) takes place.*

Theorem 2.4. *Suppose that d.f. G satisfies the conditions of the theorem 2.3. If the inequality*

$$\gamma > \frac{1}{2} \tag{2.12}$$

is true then (2.8).

In more strong conditions on d.f. G the theorem 2.3 has the following modification.

Theorem 2.5. *Suppose that the condition (2.10) is true then the following statements take place.*

1. If d.f. G has a density and there exists $\nu > 0$ so that

$$M \exp(\nu x(k, j)) < \infty, \tag{2.13}$$

then

$$\ln p_n \sim -\frac{n^{1-2\gamma}}{2\sigma^2}, \quad n \to \infty. \tag{2.14}$$

2. If there exists $\mu > 2$ so that

$$M x^\mu(k, j) < \infty, \tag{2.15}$$

then there is the positive number q_μ so that

$$p_n \le \frac{q_\mu}{n^{\mu-1-\mu\gamma}}, \quad n = 1, 2, \ldots \tag{2.16}$$

Theorems 2.1, 2.3 proves

Suppose that X_k, $k = 1, 2, \ldots$, is the sequence of independent and identically distributed r.v.'s (the sequence of i.i.d.r.v.'s), $MX_k = 0$, $p(X_k < x) = F(x)$, and for some $\alpha, 1 < \alpha < 2$, there is $C, C > 0$, and $p, q, p \ge 0, q \ge 0, p + q = 1$, so that for $x \to +\infty$ the following formulas are true:

$$1 - F(x) + F(-x) \sim \frac{C(2 - \alpha)}{\alpha} x^{-\alpha}, \tag{2.17}$$

$$\frac{1 - F(x)}{1 - F(x) + F(-x)} \to p, \quad \frac{F(-x)}{1 - F(x) + F(-x)} \to q. \tag{2.18}$$

Then accordingly to [9, chapter. 8, §9, the theorem 15 and the remark 13], [24, the chapter XVII, §5, the theorem 3] for any u, $-\infty < u < \infty$:

$$\lim_{n \to \infty} p\left(\frac{X_1 + \ldots + X_n}{n^{1/\alpha}} \ge u\right) = 1 - P(u; \alpha, C, p, q). \tag{2.19}$$

D.f. $P(u; \alpha, C, p, q)$ is stable and has the characteristic function $\varphi(t) = e^{\psi(t)}$ where

$$\psi(t) = |t|^\alpha\, C \frac{\Gamma(3 - \alpha)}{\alpha(\alpha - 1)} \left[\cos \frac{\pi\alpha}{2} \pm i(p - q) \sin \frac{\pi\alpha}{2}\right] \tag{2.20}$$

and for $p > 0$ there exists $C'(\alpha, C, p, q) > 0$ so that

$$1 - P(u; \alpha, C, p, q) \sim C'(\alpha, C, p, q)u^{-\alpha}, \quad u \to +\infty. \tag{2.21}$$

In the formula (2.20) for $t > 0$ the upper sign is "+" and for $t < 0$ the low sign is "−". More detailed information about the function $P(u; \alpha, C, p, q)$ is in [98, the chapter. 2, §7, the figure 4]. If the conditions (2.17), (2.18) are true then the formulas (2.19), (2.20), (2.21) lead to

$$\lim_{n \to \infty} P\left(X_1 + \ldots + X_n \geq 0\right) = 1 - P(0; \alpha, C, p, q) > 0. \tag{2.22}$$

Lemma 2.1. *Suppose that $F(x) = G(x+1)$ and for some $\alpha, 1 < \alpha < 2$. d.f. G satisfies the theorem 2.1 conditions then*

$$\lim_{n \to \infty} P\left(X_1 + \ldots + X_n \geq 0\right) = 1 - P(0; \alpha, C, 1, 0) > 0. \tag{2.23}$$

Proof. The condition (2.2) and the equality $F(x) = G(x+1)$ lead to the formulas (2.17), (2.18) for $p = 1$, $q = 0$. Then the formula (2.21) is true and so the formulas (2.22), (2.23) are true.

Denote

$$
\begin{aligned}
S_k &= X_1 + \ldots + X_k, \quad M_k = \max(S_1, \ldots, S_k), \\
B &= \{S_{2n} > b\}, \quad A = \{M_n > b\}, \\
A_k &= \{S_i \leq b, \quad 1 \leq i \leq k-1, \quad S_k > b\}, \\
a(\alpha) &= 1 - P(0; \alpha, C, 1, 0) > 0.
\end{aligned}
$$

Using the formula (2.23), choose $N(\alpha) > 0$ so that for $n \geq N(\alpha)$

$$P(X_1 + \ldots + X_n \geq 0) \geq \frac{a(\alpha)}{2}. \tag{2.24}$$

Lemma 2.2. *If the lemma 2.1 conditions are true for $n \geq N(\alpha)$ then*

$$P(M_n > b) \leq \frac{2}{a(\alpha)} P(S_{2n} > b). \tag{2.25}$$

Proof. Using the construction of the monograph [71] (see the proves of the lemmas 1, §4, the chapter 4) obtain for $k = 1, \ldots, n$

$$P(B \cap A_k) \geq P((S_{2n} \geq S_k) \cap A_k) = P(A_k)P(X_{k+1} + \ldots + X_{2n} \geq 0) =$$

$$= P(A_k)P(S_{2n-k} \geq 0). \tag{2.26}$$

As the condition (2.24) is true then for $k = 1, \ldots, n$

$$P(S_{2n-k} \geq 0) \geq \frac{a(\alpha)}{2}. \tag{2.27}$$

The events A_k, $k = 1, \ldots, n$, are mutually nonintersecting and so from (2.26), (2.27)

$$P(B) \geq \sum_{k=1}^{n} P(B \cap A_k) \geq \frac{a(\alpha)}{2} \sum_{k=1}^{n} P(A_k) = \frac{a(\alpha)}{2} P(A). \tag{2.28}$$

Put the events A, B into the formula (2.28), which is true for $n \geq N(\alpha)$, and obtain (2.25).

Estimate now the probability

$$A(n) = p\left(\sup_{m \geq 1}(S_{mn} - mnb) > 0\right),$$

denoting

$$C_k = \left\{\max\left(\frac{S_j}{j}, \ n2^{k-1} \leq j < n2^k\right) > b\right\}.$$

Lemma 2.3. *If the lemma 2.1 conditions and the formulas (2.24) for $n \geq N(\alpha)$ are true then*

$$A(n) \leq \sum_{k=1}^{\infty} \frac{2}{a(\alpha)} P(S_{n2^{k+1}} > nb2^{k-1}). \tag{2.29}$$

Proof. It is clear that

$$A(n) = P\left(\sup_{m \geq 1}\left(\frac{S_{mn}}{m} - nb\right) > 0\right) = P\left(\sup_{m \geq 1}\frac{S_{mn}}{m} > nb\right). \tag{2.30}$$

Using the formula (2.30) and the construction of the monograph [10, the chapter 8, §4, the theorem 5], obtain

$$A(n) \leq P\left(\sup_{m \geq 1}\frac{S_{mn}}{m} > nb\right) \leq \sum_{k=1}^{\infty} P(C_k) \leq$$

$$\sum_{k=1}^{\infty} P(M_{n2^k} > nb2^{k-1}). \tag{2.31}$$

Using the inequality (2.25), from the formula (2.31) obtain the formula (2.29).

Denote

$$\overline{F}(t) = 1 - F(t), \quad \overline{F}_n(t) = p(S_n \geq t), \quad \mu_1(y) = \int_{-y}^{y} s\, dF(s),$$

$$\gamma_t(y) = \int_{-y}^{y} |s|^t dF(s), \quad C_t = \int_{-\infty}^{\infty} |s|^t dF(s).$$

Lemma 2.4. *If the conditions of of the lemma 2.1 are true for any y, $y > 0$, t, $1 < t < \alpha$, then*

$$C_t < \infty, \quad |\mu_1(y)| \leq \frac{C_t}{y^{t-1}}, \quad \overline{F}(y) \leq \frac{C_t}{y^t}. \tag{2.32}$$

Proof. As the lemma 2.1 conditions are true then $C_t < \infty$ for $1 < t < \alpha$. It is clear that

$$\mu_1(y) = - \int_{|s| \geq y} s\, dF(s), \quad y > 0,$$

and consequently for $y > 0$

$$|\mu_1(y)| \leq \int_{|s| \geq y} |s| dF(s) = \int_{|s| \geq y} \frac{|s|^t}{|s|^{t-1}} dF(s) \leq$$

$$\leq \int_{|s| \geq y} \frac{|s|^t}{y^{t-1}} dF(s) \leq \frac{1}{y^{t-1}} \int_{-\infty}^{\infty} |s|^t dF(s) = \frac{C_t}{y^{t-1}}.$$

Analogously the inequality $\overline{F}(y) \leq C_t/y^t$ is proved.

Lemma 2.5. *If the conditions of the lemma 2.1 are true then for any τ, c, $1 < \tau < \alpha < 2$, $c > 0$, there exist $N(\tau, c)$, $Q(\tau, c)$ so that for all $n > N(\tau, c)$ the inequality*

$$\overline{F}_n(x) \leq \frac{nQ(\tau, c)}{x^\tau}, \quad x \geq cn^{1/\tau} \ln^2 n \tag{2.33}$$

is true.

Proof. Fix c, $c > 0$, and τ, $1 < \tau < \alpha$. Choose t, satisfying the inequality $1 < \tau < t < \alpha$, and use the theorem 2 from [26]. Then in conditions of the lemma 2.5 obtain

$$\overline{F}_n(x) \leq n\overline{F}(y) + \exp(Z), \quad x > 0, \quad y > 0, \quad n = 1, 2, \ldots,$$

$$Z = \frac{x}{y} - \left(\frac{x - n\mu_1(y)}{y} + \frac{n\gamma_t(y)}{y^t} \right) \ln \left(\frac{xy^{t-1}}{n\gamma_t(y)} + 1 \right). \qquad (2.34)$$

Denote $R_n(x) = nC_t \ln^t x / x^t$. The function $R_n(x)$ and the function $R_n(x)/\ln x$ monotonically decrease for $x > e$. Analogously to [53] define $y = x/\ln x$. Then accordingly to the inequality (2.32) the formula (2.34) leads to

$$\overline{F}_n(x) \le R_n(x) + \exp\left\{ \ln x - (\ln x - R_n(x)) \ln\left(1 + \frac{\ln x}{R_n(x)} \right) \right\} \le$$

$$\le R_n(x) + \exp\left\{ \ln x \left(1 - \left(1 - \frac{R_n(x)}{\ln x} \right) \ln\left(1 + \frac{\ln x}{R_n(x)} \right) \right) \right\}. \qquad (2.35)$$

Choose $N_0 > N(\alpha)$ from the condition: $cN_0^{1/t} \ln^2 N_0 > e$, where e is the foot of the natural logarithm. As the function $R_n(x)$, $x > e$, monotonically decreases by x for $n \ge N_0$ then

$$\sup\left(R_n(x),\ x \ge cn^{1/t} \ln^2 n \right) =$$

$$= R_n(cn^{1/t} \ln^2 n) = \frac{C_t(\ln c + \ln n^{1/t} + 2\ln\ln n)^t}{c^t \ln^{2t} n}.$$

It is possible to choose Q_1, $Q_1 > 0$, and N_1, $N_1 > N_0$, so that for $n \ge N_1$

$$R_n(x) \le \frac{Q_1}{\ln^t n},\ x \ge cn^{1/t} \ln^2 n. \qquad (2.36)$$

Accordingly to (2.36)

$$\inf\left(\frac{\ln x}{R_n(x)},\ x \ge cn^{1/t} \ln^2 n \right) = \frac{\ln(cn^{1/t} \ln^2 n)}{R_n(cn^{1/t} \ln^2 n)} \ge \frac{\ln^t n\ \ln(cn^{1/t} \ln^2 n)}{Q_1}.$$

So it is possible to choose $N_2 > N_1$, $Q_2 > 0$ so that for $n \ge N_2$

$$\frac{\ln x}{R_n(x)} \ge Q_2 \ln^{t+1} n = e_n,\quad x \ge cn^{1/t} \ln^2 n. \qquad (2.37)$$

Combine the formulas (2.35), (2.37) and find for $n \ge N_2$

$$\overline{F}_n(x) \le R_n(x) + \exp\{[1 - (1 - e_n^{-1})\ln(1 + e_n)]\ln x\} =$$

$$= \frac{nC_t \ln^t x}{x^t} + \exp\{[1 - (1 - e_n^{-1})\ln(1 + e_n)]\ln x\}, \qquad (2.38)$$

$$x \ge cn^{1/t} \ln^2 n.$$

The inequality (2.38) may be rewritten in the form

$$\overline{F}_n(x) \le \frac{nC_t \ln^t x}{x^t} + x^{-b_n},\qquad (2.39)$$

where

$$b_n = -1 + (1 - e_n^{-1})\ln(1 + e_n) \to \infty,\quad n \to \infty.\qquad (2.40)$$

Using the formula (2.40), choose $N_3 > N_2$ so that for $n \ge N_3$

$$b_n > 2t.\qquad (2.41)$$

Then for $n \ge N_3$, $x \ge cn^{1/t}\ln^2 n$ it is possible to rewrite the inequality (2.39) with the help of the formula (2.41) as follows

$$\overline{F}_n(x) \le \frac{nC_t \ln^t x}{x^t} + \frac{1}{x^{2t}} \le$$

$$\le \quad \frac{nC_t \ln^t x}{x^t}\left(1 + \frac{1}{C_t x^t}\right) \quad \le \quad \frac{nC_t \ln^t x}{x^t}\left(1 + \frac{1}{C_t c^t n}\right).\qquad (2.42)$$

Denote $Q_3 = C_t\left(1 + 1/C_t c^t\right)$ and obtain from the formula (2.42) for $n \ge N_3$ that

$$\overline{F}_n(x) \le \frac{nQ_3 \ln^t x}{x^t},\quad x \ge cn^{1/t}\ln^2 n.\qquad (2.43)$$

Choose Q_4 from the condition

$$\frac{\ln^t x}{x^t} < \frac{Q_4}{x^\tau},\quad x \ge e.$$

Put $N(\tau,\, c) = N_3$, $Q(\tau,\, c) = Q_3\, Q_4$. With the help of the inequality (2.43) it is possible to prove that for $n > N(\tau,\, c)$

$$\overline{F}_n(x) \le \frac{nQ(\tau,\, c)}{x^\tau},\quad x \ge cn^{1/t}\ln^2 n.$$

So for $n > N(\tau,\, c)$

$$\overline{F}_n(x) \le \frac{nQ(\tau,\, c)}{x^\tau},\quad x \ge cn^{1/\tau}\ln^2 n.$$

Lemma 2.6. *If $0 < \gamma < 1 - 1/\alpha$ then for each τ, $1/(1-\gamma) < \tau < \alpha$, it is possible to choose N'_τ, c_τ so that for $n \ge N'_\tau$, $k = 1, 2, \ldots$*

$$n^{1-\gamma}\, 2^{k-1} \ge c_\tau\, (n2^{k+1})^{1/\tau}\, \ln^2(n2^{k+1}).\qquad (2.44)$$

Proof. Choose N'_τ, c_τ from the conditions

$$\frac{n^{1-\gamma-1/\tau}}{\ln^2 n} \geq 1, \quad n \geq N'_\tau > e, \tag{2.45}$$

$$c_\tau = \min\left\{\left(2^{k(1/\tau-1)}2^{2+1/\tau}(1+(k+1)^2\ln^2 2)\right)^{-1}, k=1,2,\ldots\right\}. \tag{2.46}$$

If the formula (2.45) and the lemma 2.6 conditions are true then there exists the finite number N'_τ. The formula (2.46) leads to the inequality $c_\tau > 0$. Estimate the right side of the formula (2.44), denoting it by J. For $n \geq N'_\tau$, $k = 1, 2, \ldots$:

$$J = c_\tau\,(n2^{k+1})^{1/\tau}\,\ln^2(n2^{k+1}) = c_\tau\,(n2^{k+1})^{1/\tau}\left(\ln n + \ln 2^{k+1}\right)^2 \leq$$

$$\leq c_\tau\, n^{1/\tau}\, 2^{(k+1)/\tau}\, 2\left(\ln^2 n + (k+1)^2\ln^2 2\right) =$$

$$= c_\tau\, n^{1/\tau}\, \ln^2 n\, 2^{1+(k+1)/\tau}\left(1 + \frac{(k+1)^2\ln^2 2}{\ln^2 n}\right).$$

As $n \geq N'_\tau > e$ then

$$J \leq c_\tau\, n^{1/\tau}\, \ln^2 n\, 2^{1+(k+1)/\tau}\left(1 + (k+1)^2\ln^2 2\right).$$

Accordingly to (2.45) obtain

$$J \leq c_\tau\, n^{1-\gamma}\, 2^{k-1}\, 2^{1+(k+1)/\tau}\, 2^{1-k}\left(1 + (k+1)^2\ln^2 2\right) =$$

$$= n^{1-\gamma}\, 2^{k-1}\, c_\tau\left\{2^{k(1/\tau-1)}\, 2^{2+1/\tau}\left(1 + (k+1)^2\ln^2 2\right)\right\} \leq n^{1-\gamma}2^{k-1}.$$

The last inequality is the corollary of the formula (2.46).

Lemma 2.7. *Suppose that the conditions of the lemma 2.1 are true. If $0 < \gamma < 1 - 1/\alpha$ then for each τ, $1/(1-\gamma) < \tau < \alpha$, for $n \geq N_\tau$ and $N_\tau = \max(N'_\tau, N(\tau, c_\tau))$, $Q_\tau = Q(\tau, c_\tau)$ obtain*

$$A(n) \leq \frac{8Q_\tau}{a(\alpha)(1 - 2^{1-\tau})n^{(1-\gamma)\tau-1}}. \tag{2.47}$$

Proof. Fix τ, satisfying the inequality $1/(1-\gamma) < \tau < \alpha$. As for all $n \geq N_\tau \geq N'_\tau$, $k = 1, 2, \ldots$, the lemma 2.6 leads to the formula (2.44). So the lemma 2.5 with $c = c_\tau$ and n replaced by $n2^{k+1}$ may be applied to the inequality $P\left(S_{n2^{k+1}} > nb2^{k-1}\right)$:

$$P\left(S_{n2^{k+1}} \geq nb2^{k-1}\right) = \overline{F}_{n2^{k+1}}(n^{1-\gamma}2^{k-1}) \leq \frac{n2^{k+1}Q_\tau}{(n^{1-\gamma}2^{k-1})^\tau}, \tag{2.48}$$

$$n \geq N_\tau.$$

Put the inequality (2.48) into (2.29) and obtain for $n \geq N_\tau$:

$$A(n) \leq \sum_{k=1}^{\infty} \frac{2}{a(\alpha)} P(S_{n2^{k+1}} > nb2^{k-1}) \leq$$

$$\leq \sum_{k=1}^{\infty} \frac{2}{a(\alpha)} \frac{Q_\tau 2^{k(1-\tau)}2^{1+\tau}}{n^{(1-\gamma)\tau-1}} < \frac{8Q_\tau}{a(\alpha)(1 - 2^{1-\tau})n^{(1-\gamma)\tau-1}}.$$

The formula (2.47) is proved.

Now begin to prove the theorems 2.1, 2.3. For this aim choose

$$X_{n(k-1)+j} = x(k,j) - 1, \quad k \geq 1, \ j = 1, \ldots, n.$$

Then accordingly to the formula (2.1) and the theorems 2.1, 2.3 conditions obtain

$$p_n = A(n), \ n = 1, 2, \ldots \tag{2.49}$$

Theorem 2.1 proof. Suppose that $\gamma < 1 - 1/\alpha$. Choose arbitrary τ, satisfying the inequality $1/(1-\gamma) < \tau < \alpha$. Using the lemma 2.7 define Q_τ, N_τ so that for $n \geq N_\tau$ the inequality (2.47) is true. Put

$$C_1(\tau) - \frac{8Q_\tau}{a(\alpha)(1 - 2^{1-\tau})}.$$

Then from the formulas (2.47), (2.49) obtain the inequality (2.4).

Suppose now that $\gamma \geq 1 - 1/\alpha$ then from the equality (2.49) obtain

$$p_n \geq P(S_n > n^{1-\gamma}) = P\left(\frac{S_n}{n^{1/\alpha}} > n^{1-\gamma-1/\alpha}\right) \geq P\left(\frac{S_n}{n^{1/\alpha}} \geq 1\right), \tag{2.50}$$

$$n = 1, 2, \ldots$$

From the formulas (2.19), (2.22) find that

$$\lim_{n \to \infty} P\left(\frac{S_n}{n^{1/\alpha}} \geq 1\right) = 1 - P(1, \alpha, C, 1, 0) > 0. \tag{2.51}$$

The formulas (2.50), (2.51) lead to

$$\liminf_{n \to \infty} p_n \geq 1 - P(1, \alpha, C, 1, 0) > 0$$

so the inequality (2.5) is true. The theorem 2.1 is proved.

Theorem 2.3 proof. Accordingly to the formula (2.30) obtain

$$A(n) = P\left(\sup_{m \geq 1} \frac{S_{mn}}{m} > nb\right) \leq P\left(\sup_{m \geq 1} \left|\frac{S_{mn}}{m}\right| > nb\right). \tag{2.52}$$

Denote

$$\overline{C}_k = \left\{\max\left(\left|\frac{S_j}{j}\right|, \quad n2^{k-1} \leq j < n2^k\right) > b\right\},$$

$$\overline{M}_n = \max(|S_1|, \ldots, |S_n|),$$

and from the inequality (2.52) analogously to the formula (2.31) obtain

$$A(n) \leq \sum_{k=1}^{\infty} P(\overline{M}_{n2^k} > nb2^{k-1}). \tag{2.53}$$

Using the Kolmogorov's inequality obtain from (2.53):

$$A(n) \leq \sum_{k=1}^{\infty} \frac{n2^k \sigma^2}{(nb2^{k-1})^2}. \tag{2.54}$$

If the condition $\gamma < 1/2$ is true then from the formula (2.54) obtain

$$A(n) \leq \frac{4\sigma^2}{n^{1-2\gamma}} \to 0, \quad n \to \infty. \tag{2.55}$$

Put $C_2 = 4\sigma^2$ and obtain from the formulas (2.1), (2.49), (2.55) that

$$p_n \leq \frac{C_2}{n^{1-2\gamma}}, \quad n = 1, 2, \ldots \tag{2.56}$$

The formula (2.56) leads to the inequality (2.11).

Suppose that $\gamma \geq 1/2$ and analogously to the formula (2.50) obtain

$$p_n \geq P(S_n > n^{1-\gamma}) = P\left(\frac{S_n}{n^{1/2}} > n^{1-\gamma-1/2}\right) \geq P\left(\frac{S_n}{n^{1/2}} \geq 1\right). \tag{2.57}$$

Accordingly to the theorem 2.3 conditions and to the Lindeberg theorem corollary [30, the chapter 8, §40] obtain the equality

$$\lim_{n \to \infty} P\left(\frac{S_n}{n^{1/2}} \geq 1\right) = \int_1^{\infty} \frac{e^{-t^2/2\sigma^2} dt}{\sqrt{2\pi}\sigma} = \overline{\Phi}_{0,\sigma^2}(1) > 0, \tag{2.58}$$

where $\overline{\Phi}_{0,\sigma^2}(t)$ is the tail of the Gaussian distribution $\Phi_{0,\sigma^2}(t)$ with the mean 0 and the variance σ^2.

Then from (2.57) and (2.58) obtain the formula

$$\lim_{n\to\infty} \inf p_n \geq \overline{\Phi}_{0,\sigma^2}(1) > 0$$

and so the inequality (2.5). The theorem 2.3 is proved.

Theorems 2.2, 2.4 proves

Denote

$$y(k,j) = x(k,j) - 1, \; z_n(k) = \frac{1}{n^{1/\alpha}} \sum_{j=1}^{n} y(k,j), \; k \geq 1, \; n \geq 1.$$

Lemma 2.8. *Suppose that for some* α, $1 < \alpha < 2$, *d.f.* G *satisfies the theorem 2.1 conditions. Then for* $n \to \infty$ *the weak convergence of d.f.* $P(z_n(k) < t)$ *to d.f.* $P(t; \alpha, C, 1, 0)$ *(in all continuous points of* $P(t; \alpha, C, 1, 0)$*) is true, where* $P(t; \alpha, C, 1, 0)$ *is the stable distribution with the characteristic function (c.f.)* $\varphi(t) = e^{\psi(t)}$ *and* $\psi(t)$ *is defined in (2.20).*

Proof. This statement is the formulas (2.17) – (2.20) corollary for $F(x) = G(x + 1)$, $p - 1$, $q - 0$.

Remark 1. *Accordingly to the formula (2.20) d.f.* $P(t; \alpha, C, 1, 0)$ *has a density. The density of d.f.* $P(t; \alpha, C, 1, 0)$ *[98, the theorem 2.7.5] is bounded* $P(t; \alpha, C, 1, 0)$.

Lemma 2.9. *Suppose that d.f.* G *satisfies the theorem 2.3 conditions. Then for* $n \to \infty$ *the weak convergence of d.f.* $P(z_n(k) < t)$ *to d.f.* $\Phi_{0,\sigma^2}(t)$ *is true.*

Proof. The lemma 2.9 statement is the direct Lindeberg theorem [30, the chapter 8, §40] corollary.

Lemma 2.10. *The following equality is true:*

$$p_n = P\left(\sup\left(\sum_{k=1}^{m}\left(z_n(k) - n^{1-\gamma-1/\alpha}\right), \; m = 1, 2, \ldots\right) > 0\right). \tag{2.59}$$

Proof.

$$p_n = P\left(\sup\left(\sum_{k=1}^{m}\sum_{j=1}^{n}\left(x(k,j)-1-b\right),\quad m=1,2,\ldots\right)>0\right)=$$

$$= P\left(\sup\left(\sum_{k=1}^{m}\sum_{j=1}^{n}\left(y(k,j)-n^{-\gamma}\right),\quad m=1,2,\ldots\right)>0\right)=$$

$$= P\left(\sup\left(\sum_{k=1}^{m}\left(z_n(k)-n^{1-\gamma-1/\alpha}\right),\quad m=1,2,\ldots\right)>0\right).$$

Suppose that $z(1),z(2),\ldots$ are i.i.d.r.v.'s with the following d.f.: if α, $1<\alpha<2$, then d.f. $P(z_1<t)=G_\alpha(t)=P(t;\alpha,C,1,0)$, if $\alpha=2$ then d.f. $P(z_1<t)=G_2(t)=\Phi_{0,\sigma^2}(t)$. Put

$$Z(M)=\sup\left(\sum_{k=1}^{m}z(k),\ 1\le m\le M\right),\ Z(\infty)=Z,$$

$$Z_n(M)=\sup\left(\sum_{k=1}^{m}z_n(k),\quad 1\le m\le M\right),\ Z_n(\infty)=Z_n.$$

Lemma 2.11. *For any α, $1<\alpha\le 2$, the equality*

$$P\left(Z>1\right)=1 \tag{2.60}$$

is true.

Proof. Fix α, $1<\alpha\le 2$. By the definition $Mz(k)=0$ then for any bounded interval Δ on the straight line

$$\limsup_{m\to\infty} P\left(\sum_{k=1}^{m}z(k)\in\Delta\right)=$$

$$= \limsup_{m\to\infty} P\left(\frac{1}{m^{1/\alpha}}\sum_{k=1}^{m}z(k)\in\frac{1}{m^{1/\alpha}}\Delta\right). \tag{2.61}$$

If Δ is fixed then there exists M so that for $m\ge M$

$$\frac{1}{m^{1/\alpha}}\Delta\subset(-1,1). \tag{2.62}$$

Then accordingly to the formulas (2.61). (2.62)

$$\limsup_{m\to\infty} P\left(\sum_{k=1}^{m} z(k) \in \Delta\right) \le$$

$$\le \limsup_{m\to\infty} P\left(\frac{1}{m^{1/\alpha}} \sum_{k=1}^{m} z(k) \in (-1,1)\right). \quad (2.63)$$

As r.v.'s $z(k)$ for some α, $1 < \alpha \le 2$, have d.f. G_α then

$$P\left(\frac{1}{m^{1/\alpha}} \sum_{k=1}^{m} z(k) \in (-1,1)\right) \equiv P\left(z(1) \in (-1,1)\right).$$

So

$$\limsup_{m\to\infty} P\left(\frac{1}{m^{1/\alpha}} \sum_{k=1}^{m} z(k) \in (-1,1)\right) =$$

$$= P\left(z(1) \in (-1,1)\right) = 1 - \epsilon_1, \quad \epsilon_1 > 0. \quad (2.64)$$

The formulas (2.63), (2.64) allow to define $\epsilon_1 > 0$, satisfying for any bounded interval Δ the inequality

$$\limsup_{m\to\infty} P\left(\sum_{k=1}^{m} z(k) \in \Delta\right) \le 1 - \epsilon_1.$$

The conditions of [11, the chapter 1, §3, the theorem 7] are true and so the equality (2.60) takes place.

Lemma 2.12. *For any α, $1 < \alpha \le 2$, ϵ, $\epsilon > 0$, there exists $M'(\alpha, \epsilon)$ so that for all $M \ge M'(\alpha, \epsilon)$*

$$P\left(Z(M) > 1\right) > 1 - \epsilon.$$

Proof. Using the lemma 2.11 statement (see the formula (2.60)), find

$$1 = P\left(Z = Z(\infty) > 1\right) = \lim_{M\to\infty} P\left(Z(M) > 1\right).$$

Introduce the Markov chains $w_n(m)$, $m \ge 1$, $w(m)$, $m \ge 1$, by the formulas

$$w_n(0) = 0, \; w_n(m+1) = \max(w_n(m) + z_n(m+1), 0), \quad (2.65)$$

$$w(0) = 0, \; w(m+1) = \max(w(m) + z(m+1), 0), \; m = 1, 2, \dots \quad (2.66)$$

From [11, the chapter 1, §3, the theorem 2] obtain the coincidence of r.v.'s $Z_n(m)$ and $w_n(m)$, $m = 1, 2, \dots$, by the distribution. Analogously r.v.'s $Z(m)$ and $w(m)$, $m = 1, 2, \dots$, coincide by the distribution too.

Lemma 2.13. *For any α, $1 < \alpha \leq 2$, and for any $m > 0$ if $n \to \infty$ then there is the weak convergence of r.v.'s $w_n(m)$ distributions to r.v. $w(m)$ distribution (and so for r.v.'s $Z_n(m)$ distributions to r.v. $Z(m)$ distribution).*

Proof. From the lemmas 2.8, 2.9 obtain that if $n \to \infty$ so r.v.'s $z_n(k)$ distributions converge weakly to r.v. $z(k)$ distribution for α, $1 < \alpha \leq 2$. As the result r.v.'s $w_n(1)$ distributions converge weakly to r.v. $w(1)$ distribution for $n \to \infty$.

Suppose that r.v.'s $w_n(m)$ distributions converge weakly to r.v. $w(m)$ distribution for $n \to \infty$. Then from [64, the theorem of §7] obtain that c.f.'s of r.v.'s $w_n(m)$ converge uniformly to c.f. of r.v. $w(m)$ at each finite interval for $n \to \infty$.

As for $n \to \infty$ r.v.'s $z_n(k)$ distributions converge weakly to r.v. $z(k)$ distribution so c.f. of r.v.'s $z_n(m+1)$ converge to c.f. of r.v. $z(m+1)$ for $n \to \infty$ uniformly for at any finite interval. Consequently for $n \to \infty$ c.f.'s of r.v.'s $w_n(m) + z_n(m+1)$ converge to c.f. of r.v. $w(m) + z(m+1)$ uniformly at any finite interval. So if $n \to \infty$ then there is weak convergence of r.v.'s $w_n(m) + z_n(m+1)$ distributions to r.v. $w(m) + z(m+1)$ distribution. As the result there is the weak convergence of r.v.'s $w_n(m+1)$ distributions to r.v. $w(m+1)$ distribution for $n \to \infty$. The induction statement is proved.

Lemma 2.14. *For any α, $1 < \alpha \leq 2$, and any $m > 0$ there exist the nonnegative numbers $p(m)$, $q(m)$: $p(m) + q(m) = 1$ and d.f. $F_m(t)$ with bounded density $f_m(t)$, $-\infty < t < \infty$, $f_m(t) = 0$, $t \leq 0$, so that*

$$P\left(w(m) < t\right) = p(m)\theta(t) + q(m)F_m(t), \quad -\infty < t < \infty. \tag{2.67}$$

Here $\theta(t) = 0$, $t \leq 0$, $\theta(t) = 1$, $t > 0$.

Proof. Denote

$$g_\alpha(t) = \frac{d}{dt}P\left(z(1) < t\right) = \frac{d}{dt}G_\alpha(t).$$

As the lemmas 2.8 and 2.9 are true so for $1 < \alpha \leq 2$ the function $g_\alpha(t)$ is unimodal (that is this function possesses single local and so global extremum – maximum). Consequently $g_\alpha(t)$ has a finite upper bound at $(-\infty, \infty)$. Choose

$$p(1) = \int_{-\infty}^{0} g_\alpha(\tau)d\tau,$$

$$q(1) = 1 - p(1), \quad f_1(t) = 0, \ t \leq 0, \ f_1(t) = \frac{g_\alpha(t)}{q(1)}, \ t > 0, \tag{2.68}$$

and denote by $F_1(t)$ d.f. with the density $f_1(t)$. Then

$$P\left(w(1) < t\right) = P\left(\max(0, z(1)) < t\right) =$$

$$= \theta(t)G_\alpha(t) = p(1)\theta(t) + q(1)F_1(t) \quad (2.69)$$

and the density $f_1(t)$ is bounded.

Suppose that the representation (2.67) is true for $m = k$ and for d.f. $F_k(t)$ with bounded density $f_k(t)$. Prove this representation for $m = k + 1$. Denote by " $*$ " the operation of d.f. conjuncture. From the equality (2.67), which is true for $m = k$, obtain

$$P\left(w(k) + z(k+1) < t\right) = (p(k)\theta(t) + q(k)F_k(t)) * G_\alpha(t) =$$

$$= p(k)G_\alpha(t) + q(k)F_k(t) * G_\alpha(t) = p(k)G_\alpha(t) + q(k)R_k(t),$$

$$R_k(t) = F_k(t) * G_\alpha(t),$$

where d.f. $R_k(t)$ has bounded density

$$r_k(t) = \int_{-\infty}^{\infty} f_k(t - \tau)g_\alpha(\tau)d\tau.$$

It is clear that the density $\psi_k(t)$ of the distribution

$$P\left(w(k) + z(k+1) < t\right) = \Psi_k(t)$$

is the bounded function and $\psi_k(t) = p(k)g_\alpha(t) + q(k)r_k(t)$. Analogously to (2.68) choose

$$p(k + 1) = \int_{-\infty}^{0} \psi_k(\tau)d\tau, \quad q(k + 1) = 1 - p(k + 1),$$

$$f_{k+1}(t) = 0, \ t \le 0, \ f_{k+1}(t) = \frac{\psi_k(t)}{q(k + 1)}, \ t > 0,$$

and denote by $F_{k+1}(t)$ d.f. with the bounded density $f_{k+1}(t)$. Then

$$P\left(w(k + 1) < t\right) = P\left(\max(0, w(k) + z(k+1)) < t\right) =$$

$$= \theta(t)P\left(w(k) + z(k+1) < t\right) = \theta(t)\Psi_k(t) = p_{k+1}\theta(t) + q_{k+1}F_{k+1}(t).$$

Consequently for $m = k + 1$ the representation (2.67) is true too.

Lemma 2.15. *For any α, $1 < \alpha \leq 2$, and for any $m > 0$*

$$\lim_{n \to \infty} P\left(Z_n(m) > 1\right) = P\left(Z(m) > 1\right).$$

Proof. The equalities (2.67) and the lemma 2.13 lead that in each continuity point $t = T$ of d.f. $P\left(w(m) < t\right)$

$$\lim_{n \to \infty} P\left(Z_n(m) > T\right) = P\left(Z(m) > T\right).$$

As the lemma 2.14 is true so the point $T = 1$ is continuity point of d.f. $P\left(w(m) < t\right)$.

Lemma 2.16. *For any α, $1 < \alpha \leq 2$, and for any $\gamma > 1 - 1/\alpha$*

$$\lim_{n \to \infty} p_n = 1.$$

Proof. Suppose that $1 < \alpha \leq 2$, $\epsilon > 0$. Define by the lemma 2.12 $M' = M'(\alpha, \epsilon)$ so that

$$P\left(Z(M') > 1\right) > 1 - \epsilon. \tag{2.70}$$

Using the lemma 2.15 for fixed $M' = M'(\alpha, \epsilon)$, α, ϵ, find $N_1 = N_1(\alpha, \epsilon)$ so that for any $n \geq N_1$

$$\mid P\left(Z_n(M') > 1\right) - P\left(Z(M') > 1\right) \mid < \epsilon. \tag{2.71}$$

From the inequalities (2.70), (2.71) find that for any $n \geq N_1$

$$P\left(Z_n(M') > 1\right) > 1 - 2\epsilon. \tag{2.72}$$

The lemma 2.10 (see the formula (2.59)) leads to

$$1 \geq p_n = P\left(\sup\left(\sum_{k=1}^{m}\left(z_n(k) - n^{1-\gamma-1/\alpha}\right), \quad m = 1, 2, \ldots\right) > 0\right) \geq$$

$$\geq P\left(\sup\left(\sum_{k=1}^{m}\left(z_n(k) - n^{1-\gamma-1/\alpha}\right), \quad 1 \leq m \leq M'\right) > 0\right) \geq$$

$$\geq P\left(\sup\left(\sum_{k=1}^{m}(z_n(k)), \quad 1 \leq m \leq M'\right) - M'n^{1-\gamma-1/\alpha} > 0\right) \geq$$

$$\geq P\left(Z_n(M') > M'n^{1-\gamma-1/\alpha}\right). \tag{2.73}$$

Choose $N_2 = N_2(\alpha, \epsilon)$ so that $M'N_2^{1-\gamma-1/\alpha} < 1$. Then for $n \geq \max(N_1(\alpha, \epsilon), N_2(\alpha, \epsilon))$ from the formulas (2.72), (2.73) obtain

$$1 \geq p_n \geq P\left(Z_n(M') > 1\right) > 1 - 2\epsilon.$$

The lemma 2.16 leads to the theorems 2.2. 2.4 statements.

Theorem 2.5 proof

From the definition of p_n obtain

$$P\left(\sum_{j=1}^{n}(x(1,j)-1) > nb\right) \le p_n \le$$

$$\le \sum_{m=1}^{\infty} P\left(\sum_{k=1}^{m}\sum_{j=1}^{n}(x(k,j)-1) > mnb\right). \quad (2.74)$$

Denote $y_j = x(1,j)-1$, $j \ge 1$ and put $S_n = \sum_{j=1}^{n} y_j$, $n = 1,2,\ldots$ Rewrite the inequality (2.74) as follows

$$P(S_n > nb) \le p_n \le \sum_{m=1}^{\infty} P(S_{mn} > mnb). \quad (2.75)$$

Prove the theorem 2.5, using the inequality (2.75), in two steps.

The step 1. To estimate the probability $p(S_k > kb)$ in the conditions (2.13) use the Cramer theorem [15] with the remained member in the Petrov form [59]. **Theorem *.** *If* $x \ge 0$, $x = o(\sqrt{k})$ *and the condition (2.13) is true then*

$$p(S_k>\sigma x\sqrt{k})=\overline{\Phi}_{0,1}(x)\exp\left(\frac{x^3}{\sqrt{k}}\lambda\left(\frac{x}{\sqrt{k}}\right)\right)\left[1+O\left(\frac{x+1}{\sqrt{k}}\right)\right], \quad (2.76)$$

where

$$\lambda(t) = \sum_{j=0}^{\infty} a_j t^j.$$

Here the row $\lambda(t)$ with the coefficients calculated via

$$\gamma_k = \frac{1}{i^k}\left(\frac{d^k}{dt^k}\ln E\exp(ity_1)\right)_{t=0},$$

where i is imagine unit and the symbol $\ln$ *denotes main meaning of the logarithm so that*

$$(\ln E\exp(ity_1))_{t=0} = 0.$$

The function $\overline{\Phi}_{0,1}(x)$ may be represented in the Feller form [23]:

$$\overline{\Phi}_{0,1}(x) = \frac{e^{-x^2/2}}{x\sqrt{2\pi}}\left(1 - \frac{v(x)}{x^2}\right), \quad x > 1, \quad (2.77)$$

where $v(x)$ is some function, satisfying the inequality $0 \le v(x) \le 1$, $x > 1$.

Suppose that the inequality $\gamma < 1/2$ is true. Introduce auxiliary designations

$$1 - \frac{v(n^{-\gamma}\sqrt{n}\sigma^{-1})}{n^{1-2\gamma}\sigma^{-2}} = V_n, \ 1 + O\left(\frac{n^{-\gamma}\sqrt{nm}\sigma^{-1} + 1}{\sqrt{nm}}\right) = W_{n,m},$$

$$\exp\left(\frac{-n^{1-2\gamma}m}{2\sigma^2} + \frac{n^{1-3\gamma}m}{\sigma^3}\lambda\left(\frac{n^{-\gamma}}{\sigma}\right)\right) =$$

$$= \exp\left(\frac{-n^{1-2\gamma}m}{2\sigma^2}\left(1 - \frac{2n^{-\gamma}}{\sigma}\lambda\left(\frac{n^{-\gamma}}{\sigma}\right)\right)\right) = U_{n,m}.$$

Then the inequality (2.75) may be rewritten with the help of the formulas (2.76), (2.77) as follows

$$\frac{\sigma}{n^{-\gamma}\sqrt{2\pi n}}V_n W_{n,1} U_{n,1} \le p_n \le \sum_{m=1}^{\infty} \frac{\sigma}{n^{-\gamma}\sqrt{2\pi nm}}W_{n,m} U_{n,m}. \qquad (2.78)$$

Choose $c > 0$ so that

$$\begin{aligned} W_{n,m} &\le (1 + cn^{-\gamma}), \ m = 1, 2, \ldots, \\ W_{n,1} &\ge (1 - cn^{-\gamma}), \ n = 1, 2, \ldots \end{aligned} \qquad (2.79)$$

From the inequalities (2.78), (2.79) obtain

$$\frac{\sigma}{n^{-\gamma}\sqrt{2\pi n}}V_n(1 - cn^{-\gamma})U_{n,1} \le p_n \le$$

$$\le \frac{\sigma(1 + cn^{-\gamma})U_{n,1}}{(1 - U_{n,1})n^{-\gamma}\sqrt{2\pi n}}, \ n \ge 1. \qquad (2.80)$$

Then

$$\ln p_n \sim -\frac{n^{1-2\gamma}}{2\sigma^2}, \ n \to \infty,$$

that is

$$\lim_{n\to\infty} p_n = 0. \qquad (2.81)$$

Suppose that $\gamma \ge 1/2$ then from the inequality (2.75) obtain

$$p_n \ge P(S_n \ge n^{1-\gamma}) \ge P(S_n \ge \sqrt{n}). \qquad (2.82)$$

Using the central limit theorem [30, the Lindeberg theorem corollary] obtain from (2.82)

$$\liminf_{n\to\infty} p_n \ge \overline{\Phi}_{0,1}(1) > 0.$$

So we prove that the equality (2.81) is true if and only if the inequality $\gamma < 1/2$ takes place.

Step 2. To estimate the probability $p(S_k > kb)$ in the conditions (2.15) use the Nagaev inequality [54] in the following theorem ** form.

Theorem **. *If the conditions (2.15) are true then there exist the positive and finite numbers n_μ, g_μ so that for $x \geq n_\mu \sqrt{k}$*

$$p(S_k > x) \leq \frac{kg_\mu}{x^\mu}, \quad k = 1, 2, \ldots \tag{2.83}$$

The statement of the theorem ** leads that for

$$n \geq N_\mu = n_\mu^\beta, \ \beta = \frac{2}{1 - 2\gamma}$$

the following inequalities are true

$$p(S_{mn} > mnb) \leq \frac{mng_\mu}{(mnb)^\mu} = \frac{g_\mu}{b^\mu (mn)^{\mu-1}}, \quad m = 1, 2, \ldots \tag{2.84}$$

Using the inequalities (2.75), (2.84) obtain for $n \geq N_\mu$:

$$p_n \leq \frac{q_\mu}{b^\mu n^{\mu-1}} = \frac{q_\mu}{n^{\mu-1-\mu\gamma}}, \ n \geq 1, \ \ q_\mu = g_\mu \sum_{m=1}^{\infty} m^{1-\mu} < \infty.$$

So in this case the equality (2.81) is true if and only if $\gamma < 1/2$. The theorem 2.5 is proved.

Remark 2. *The analyzed risk model may be improved by a consideration of finite horizon ruin probabilities. From an one side it allows to investigate as a possibility so a necessity of current time coalitions caused by some short time factors. From an another side this suggestion allows to simplify the model analysis.*

Applications to queueing systems with group serving

The systems with a group serving are widely used in a modelling of computer and information transmission networks and so on. An introduction of the group serving into the oneserver queueing system demands a consideration of general d.f. of serving times and inter-arrival intervals. So we are to analyze the system $GI|GI|1|\infty$, which direct analytical investigation is sufficiently complicated task [11].

Before an application of the Monte-Carlo methods [12] it is worthy to use some asymptotic methods [8]. The group serving systems give a possibility to choose a convenient asymptotic when the group size n tends to the infinity. The statement that in the system $GI|GI|1|\infty$ with the group serving waiting time converges to zero as $n \to \infty$ is proved. This result is some modification of the theorem 2.5.

Suppose that the recurrent input flow of the initial system $GI|GI|1|\infty$ is characterized by the random intervals b_i between the arrivals of the i-th and the i-th customers and denote by a_i, $i \geq 1$, the serving times of the customers. Suppose that r.v.'s $c_i = a_i - b_i$, $i \geq 1$, are independent and identically distributed and satisfy the inequality

$$Mc_i < 0. \tag{2.85}$$

Describe the queueing system with the serving of n customers groups, denoting by

$$A_j^{(n)} = \sum_{i=n(j-1)+1}^{nj} a_i, \quad B_j^{(n)} = \sum_{i=nj}^{n(j+1)-1} b_i$$

the time of the j-th group serving and the interval between the j-th and the $(j+1)$-th groups arrivals correspondingly. It is clear that the sequence $C_j^{(n)} = A_j^{(n)} - B_j^{(n)}$, $j \geq 1$, consists of i.i.d.r.v.'s and from (2.85)

$$MC_j^{(n)} = nMc_i < 0. \tag{2.86}$$

Suppose that $W_j^{(n)}$ is the time between the formation of the j-th group and the begin of its serving, then

$$W_{j+1}^{(n)} = \max\left(0, W_j^{(n)} + C_j^{(n)}\right), \quad j \geq 1. \tag{2.87}$$

Suppose that $W_1^{(n)} = 0$ then we obtain the Markov chain $W_j^{(n)}$, $j \geq 1$. As the formulas (2.85), (2.86) are true so this chain has the limit distribution

$$\lim_{j \to \infty} P\{W_j^{(n)} > t\} = P\{W^{(n)} > t\}, \quad t \geq 0,$$

$$W^{(n)} = \sup\left\{0, \sum_{j=1}^{k} C_j^{(n)}, \ k \geq 1\right\}. \tag{2.88}$$

The sequence $P\{W_j^{(n)} > t\}$, $j \geq 1$, is monotonically increasing by j.

Theorem 2.6. *1) Suppose that for some $\mu > 0$*

$$Me^{\mu a_i} < 0 \tag{2.89}$$

(Cramer condition). Then there are the positive numbers $c, d,\ d<1,$ so that

$$P\{W^{(n)} > 0\} \leq cd^n, \quad n \geq 1. \tag{2.90}$$

2) If the Cramer condition is replaced by the following condition: for some $m > 2$

$$Ma_i^m < \infty, \tag{2.91}$$

then there exists the positive number q_m so that

$$P\{W^{(n)} > 0\} \leq \frac{q_m}{n^{m-1}}, \quad n \geq 1. \tag{2.92}$$

Proof. 1) The inequalities (2.85), (2.89) allow to find the positive number ν so that

$$Me^{\nu c_i} = f < 1. \tag{2.93}$$

Denote $S_k^{(n)} = \sum_{j=1}^{k} C_j^{(n)}$ then the formulas, (2.88), (2.93) lead to

$$P\{W^{(n)} > 0\} = P\{\sup\{0, S_k^{(n)},\ k \geq 1\} > 0\} \leq \sum_{k=1}^{\infty} P\{S_k^{(n)} > 0\} \leq$$

$$\leq \sum_{k=1}^{\infty} Me^{\nu S_k^{(n)}} \leq \sum_{k=1}^{\infty} f^{kn} = \frac{f^n}{1 - f^n}, \quad n \geq 1. \tag{2.94}$$

So for $c = 1/(1 - f^n)$, $d = f$ we obtain the inequality (2.90).

2) Denote $c_i' = a_i - b_{n+i}$ then the equality

$$S_k^{(n)} = \sum_{i=1}^{nk} c_i' = \sum_{i=1}^{nk} \Delta c_i - nkc, \tag{2.95}$$

with $c = -Mc_i > 0$, $\Delta c_i = c_i' + c$ is true. From the condition (2.91) obtain

$$M(\max(0, \Delta c_i))^m < \infty. \tag{2.96}$$

Analogously with the formula (2.94)

$$P\{W^{(n)} > 0\} \leq \sum_{k=1}^{\infty} P\{S_k^{(n)} > 0\} = \sum_{k=1}^{\infty} P\left\{\sum_{i=1}^{nk} \Delta c_i > nkc\right\}. \tag{2.97}$$

Using the condition (2.96) and the theorem 2.5 proof it is possible to find n_m, g_m so that for $n \geq n_m$, $k \geq 1$

$$P\left\{ \sum_{i=1}^{nk} \Delta c_i > nkc \right\} \leq \frac{2nkg_m}{(nkc)^m} = \frac{2g_m}{c^m(nk)^{m-1}}. \qquad (2.98)$$

Combining the formulas (2.97),(2.98), obtain

$$P\{W^{(n)} > 0\} \leq \frac{q_m}{n^{m-1}}, \quad n \geq 1, \quad \text{where} \quad q_m = \frac{2g_m}{c^m} \sum_{k=1}^{\infty} \frac{1}{k^{m-1}} < \infty.$$

So the inequality (2.92) and consequently the theorem 2.6 are proved.

Remark 3. *The theorem 2.6 allows in the asymptotic $n \to \infty$ to replace considered system $GI|GI|1|\infty$ with group serving by the system $GI|GI|\infty$. It is the cooperative effect in the group serving.*

2.2.   Mutual insurance models with weak dependent risks

In the previous subsection the ruin probability Φ of the aggregated insurance system, consisted of n subsystems with independent and identically distributed annual risks, was considered. In different conditions the parameter $\gamma^* > 0$, satisfying

$$\Phi = \left\{ \begin{array}{ll} 0, & \gamma < \gamma^*, \\ 1, & \gamma > \gamma^*. \end{array} \right. \qquad (2.99)$$

was found.

P. Embrechts suggested to consider the phase transition (2.99) in the case when the risks of different united subsystems are weak dependent. In this subsection there is an exhaustive solution of this question, based on a special stochastic model of a weak dependence between annual risks $x(k, j)$ of aggregated subsystems. This dependence supposed that the fluctuation of the risks $x(k, j)$ is divided into a common part with the small order $n^{-\delta}$, $\delta > 0$, and an individual part with the finite order 1 as follows

$$x(k, j) - 1 = n^{-\delta}\Delta x(k) + \Delta x(k, j). \qquad (2.100)$$

Here $\Delta x(k)$, $\Delta x(k, j)$ are independent and identically distributed r.v.'s with common d.f. $U(t)$, $E\Delta x(k) = E\Delta x(k, j) = 0$, $k \geq 1$, $1 \leq j \leq n$. As in the case of the independent risks the phase transition phenomenon is recognized. This phenomenon is showed graphically at the figure 6:

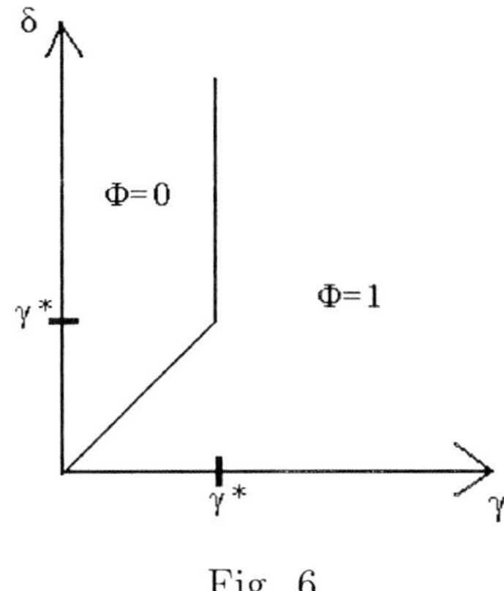

Fig. 6

Begin to make the accuracy formulation of these results.

Theorem 2.7. *Suppose that d.f.* $U(t)$, $U(-1/2) = 0$, *has bounded density and* $D\Delta x(k) = D\Delta x(k, j) = \sigma^2$, $0 < \sigma^2 < \infty$, $k \geq 1$, $1 \leq j \leq n$. *If* $0 < \gamma < 1/2$ *and* $\delta > \gamma$ *then*

$$\lim_{n \to \infty} p_n = 0. \tag{2.101}$$

If $\gamma > 1/2$ *or* $0 < \delta < \gamma$ *then*

$$\lim_{n \to \infty} p_n = 1. \tag{2.102}$$

Proof. The following formula is true:

$$p_n = P\left(\sup_{m>0}\left(\sum_{k=1}^{m}\sum_{j=1}^{n}(n^{-\delta}\Delta x(k) + \Delta x(k,j) - n^{-\gamma})\right) > 0\right) =$$

$$= P\left(\sup_{m>0}\left(\frac{n^{1-\delta}}{m}\sum_{k=1}^{m}\Delta x(k) + \frac{1}{m}\sum_{k=1}^{m}\sum_{j=1}^{n}\Delta x(k,j)\right) > n^{1-\gamma}\right) \leq$$

$$\leq P\left(\sup_{m>0}\frac{n^{1-\delta}}{m}\sum_{k=1}^{m}\Delta x(k)+\right.$$

$$\left. + \sup_{m>0}\frac{1}{m}\sum_{k=1}^{m}\sum_{j=1}^{n}\Delta x(k,j) > n^{1-\gamma}\right). \tag{2.103}$$

Then from (2.103) obtain the inequality

$$p_n \leq P_1(n) + P_2(n), \tag{2.104}$$

in which

$$P_1(n) = P\left(\sup_{m>0} \frac{n^{1-\delta}}{m} \sum_{k=1}^{m} \Delta x(k) > n^{1-\gamma}/2\right),$$

$$P_2(n) = P\left(\sup_{m>0} \frac{1}{m} \sum_{k=1}^{m} \sum_{j=1}^{n} \Delta x(k,j) > n^{1-\gamma}/2\right).$$

Denote

$$M_t = \max(|\Delta x(1)|, |\Delta x(1) + \Delta x(2)|, \ldots,$$

$$|\Delta x(1) + \Delta x(2) + \ldots + \Delta x(t)|). \tag{2.105}$$

Suppose now that $0 < \gamma < 1/2$ and $\delta > \gamma$. Then, using the algorithm of the theorem 2.1 proof obtain

$$P_1(n) = P\left(\sup_{m>0} \frac{1}{m} \sum_{k=1}^{m} \Delta x(k) > n^{\delta-\gamma}/2\right) \leq$$

$$\leq \sum_{k=1}^{\infty} P\left(\max\left(\frac{|\sum_{i=1}^{j} \Delta x(i)|}{j}, 2^{k-1} \leq j < 2^k\right) > n^{\delta-\gamma}/2\right) \leq$$

$$\leq \sum_{k=1}^{\infty} P\left(M_{2^k} > n^{\delta-\gamma} 2^{k-2}\right) \leq \sum_{k=1}^{\infty} \frac{2^k D\Delta x(k)}{(n^{\delta-\gamma} 2^{k-2})^2} = \frac{16\sigma^2}{n^{2\delta-2\gamma}}, \tag{2.106}$$

$$P_2(n) = P\left(\sup_{m>0} \frac{1}{mn} \sum_{k=1}^{m} \sum_{j=1}^{n} \Delta x(k,j) > n^{-\gamma}/2\right) \leq \frac{16\sigma^2}{n^{1-2\gamma}}. \tag{2.107}$$

The formulas (2.104), (2.106), (2.107) lead to the equality (2.101).

Prove now the equality (2.102). For this aim, divide the set $G = \{(\gamma,\delta) : 0 < \delta < \gamma \text{ or } \gamma > 1/2\}$ into the nonintersecting subsets $G_1 = \{(\gamma,\delta) : \delta \geq 1/2 \text{ and } \gamma > 1/2\}$ and $G_2 = \{(\gamma,\delta) : 0<\delta<\gamma, \ \delta < 1/2\}$, $G_1 \cup G_2 = G$. Consider the case when $(\gamma,\delta) \in G_1$. From (2.103) obtain that $\forall M, M \in N = \{1,2,\ldots\}$,

$$p_n = P\left(\sup_{m>0} \sum_{k=1}^{m} \left(n^{1/2-\delta} \Delta x(k) + \sum_{j=1}^{n} \frac{\Delta x(k,j)}{\sqrt{n}} - n^{1/2-\gamma}\right) > 0\right) \geq$$

$$\geq P\left(\max_{0<m\leq M}\sum_{k=1}^{m}\left(n^{1/2-\delta}\Delta x(k)+\sum_{j=1}^{n}\frac{\Delta x(k,j)}{\sqrt{n}}\right)>Mn^{1/2-\gamma}\right).$$

As $\forall \epsilon > 0\ \exists N_1:\ \forall n\geq N_1\ n^{1/2-\gamma}M<1$ so $\forall n\geq N_1$

$$p_n\geq P\left(\max_{0<m\leq M}\sum_{k=1}^{m}\left(n^{1/2-\delta}\Delta x(k)+\sum_{j=1}^{n}\frac{\Delta x(k,j)}{\sqrt{n}}\right)>1\right). \tag{2.108}$$

Suppose that $\lambda(s)$, $s\geq 1$, is s.i.i.d.r.v.'s with the common Gaussian d.f. $\Phi_{0,\sigma^2}(t)$, which has the mean 0 and the variance $\sigma^2(t)$. Denote

$$\lambda_{n,b}(s)=\sum_{j=1}^{n}\frac{\Delta x(s,j)}{n^{1/b}},\ u_{n,b}(s)=n^{1-1/b-\delta}\Delta x(s),$$

$$z_{n,b}(s)=u_{n,b}(s)+\lambda_{n,b}(s),\ z_b(s)=u_{n,b}(s)+\lambda(s),$$

$$Z_{n,b}(s)=\max\left(0,\max_{0<m\leq s}\sum_{k=1}^{m}z_{n,b}(k)\right),$$

$$Z_b(s)=\max\left(0,\max_{0<m\leq s}\sum_{k=1}^{m}z_b(k)\right),$$

$$w_{n,b}(s)=\max\left(0,w_{n,b}(s-1)+z_{n,b}(s)\right),$$

$$w_b(s)=\max\left(0,w_b(s-1)+z_b(s)\right),\ \ s\geq 1,$$

$$w_{n,b}(0)=w_b(0)=0. \tag{2.109}$$

In our case $b=2$ so from the lemma 2.9 obtain

$$F_{\lambda_{n,2}(s)}(t)=P(\lambda_{n,2}(s)<t)\Rightarrow\Phi_{0,\sigma^2}(t),\ \ n\to\infty, \tag{2.110}$$

where " $\Rightarrow$ " means the weak convergence of d.f. As the formulas

$$P(u_{n,2}(s)<t)=U(t),\ \text{if}\ \delta=1/2,$$

$$P(u_{n,2}(s)<t)\Rightarrow I(t),\ n\to\infty,\ \text{if}\ \delta>1/2$$

with

$$I(t)=\left\{\begin{array}{ll}1,&t>0,\\0,&t\leq 0,\end{array}\right. \tag{2.111}$$

are true so from the continuity theorem [10, the chapter 7, § 3] obtain

$$P(z_{n,2}(s)<t)=(U*F_{\lambda_{n,2}(s)})(t)\Rightarrow(U*\Phi_{0,\sigma^2})(t),\ \ n\to\infty, \tag{2.112}$$

if $\delta = 1/2$,

$$P(z_{n,2}(s) < t) \Rightarrow (I * F_{\lambda_{n,2}(s)})(t) = F_{\lambda_{n,2}(s)}(t), \quad n \to \infty, \tag{2.113}$$

if $\delta > 1/2$.

Then accordingly to [24, the tom 2, the chapter 6, § 9] the following equalities

$$P(w_{n,2}(s) < t) = P(Z_{n,2}(s) < t),$$

$$P(w_2(s) < t) = P(Z_2(s) < t), \ s \geq 1, \tag{2.114}$$

are true and from the formulas (2.110), (2.113)

$$P(z_{n,2} < t) \Rightarrow \Phi_{0,\sigma^2}.$$

So the lemma 2.13 leads to

$$P(Z_{n,2}(s) < t) \Rightarrow P(Z_2(s) < t), \quad n \to \infty. \tag{2.115}$$

The condition, that d.f. $U(t)$ density is bounded and so d.f. $(U * \Phi_{0,\sigma^2})(t)$ is bounded too, is necessary to obtain from the lemma 2.14 the following corollary. D.f. $F_{w_2(s)}(t) = P(w_2(s) < t)$ is continuous in the point $t = 1$. So from the formula (2.115) and from the lemma 2.15 obtain that for $\forall \epsilon > 0 \ \exists N_2 \in N : \ \forall n \geq N_2$ the inequality

$$P(Z_{n,2}(s) > 1) > P(Z_2(s) > 1) - \epsilon$$

is true. Then for $\forall n \geq \max(N_1, N_2)$ obtain

$$p_n \geq P\left(\max_{0 < m \leq M} \sum_{k=1}^{m} z_{n,2}(k) > 1\right) =$$

$$= P(Z_{n,2}(M) > 1) > P(Z_2(M) > 1) - \epsilon.$$

The lemma 2.12 leads to

$$P(Z_2(s) > 1) \to 1, \quad M \to \infty.$$

Consequently $\exists M^* \in N : \ \forall M \geq M^*$ so that

$$p_n > 1 - 2\epsilon,$$

and then the equality (2.102) is true too.

Consider now the case $(\gamma, \delta) \in G_2$. In this case analogously to the formula (2.108) obtain that for $\forall \epsilon > 0 \; \exists N : \; \forall n \geq N$

$$p_n \geq P\left(\max_{0 < m \leq M} \sum_{k=1}^{m} (\Delta x(k) + n^{\delta - 1/2} \lambda_{n,2}(k)) > 1\right).$$

It is clear that

$$P(n^{\delta - 1/2} \lambda_{n,2}(s) < t) \Rightarrow I(t), \;\; n \to \infty,$$

and consequently

$$P(\Delta x(k) + n^{\delta - 1/2} \lambda_{n,2}(s) < t) \Rightarrow (I \star U)(t) = U(t), \;\; n \to \infty.$$

Then, introducing Markov chains

$$w'_{n,2}(s) = \max\left(0, w'_{n,2}(s-1) + \Delta x(s) + n^{\delta - 1/2} \lambda_{n,2}(s)\right),$$

$$w'_2(s) = \max\left(0, w'_2(s-1) + \Delta x(s)\right), \;\; s \geq 1, \;\; w'_{n,2}(0) = w'_2(0) = 0$$

and repeating the word by the word the proof in previous case, obtain the equality (2.102).

Theorem 2.8. *Suppose that d.f. $U(t)$, $U(-1/2) = 0$, has bounded density and for some α, C so that $1 < \alpha < 2$, $C > 0$*

$$1 - U(t) \sim \frac{C(2 - \alpha)}{\alpha} t^{-\alpha}, \; t \to \infty. \tag{2.116}$$

If $0 < \gamma < 1 - 1/\alpha$ and $\delta > \gamma$ then (2.101). If $\gamma > 1 - 1/\alpha$ or $0 < \delta < \gamma$ then the equality (2.102) is true.

Proof. Consider the case $\gamma < 1 - 1/\alpha$, $\delta > \gamma$. From the theorem 2.1 obtain that for $\forall \tau, \; 1/(1 - \gamma) < \tau < \alpha$,

$$\exists\, C_1(\tau) > 0 \; : \; P_2(n) \leq C_1(\tau) n^{1 - \tau(1 - \gamma)} 2^{\tau}.$$

The multiplier 2^{τ} occurs because here we consider the probability of the inequality

$$\sup_{m > 0} \frac{1}{mn} \sum_{k=1}^{m} \sum_{j=1}^{n} \Delta x(k, j) > n^{-\gamma}/2,$$

but not the probability of the inequality

$$\sup_{m>0} \frac{1}{mn} \sum_{k=1}^{m} \sum_{j=1}^{n} \Delta x(k.j) > n^{-\gamma},$$

as it was made in the proof of the theorem 2.1.

Consequently the equality

$$\lim_{n\to\infty} P_2(n) = 0$$

is true. As the equality (2.104) takes place so we are to show only that

$$\lim_{n\to\infty} P_1(n) = 0. \tag{2.117}$$

In the proof of previous theorem it was shown that in the case $0 < \gamma < 1/2$ and $\delta > \gamma$

$$P_1(n) \leq \sum_{k=1}^{\infty} P(M_{2^k} > n^{\delta-\gamma}2^{k-2}) = A. \tag{2.118}$$

Choose $K^* > 0 :\ K^* \geq \max(\log_2 N(\tau, 1) - 1, \log_2 N(\alpha))$, where $N(\alpha),\ N(\alpha) > 0$, is so that $\forall t \geq N(\alpha)$ (see the formula (2.24))

$$P(\Delta x(1) + \Delta x(2) + \ldots + \Delta x(t) > 0) \geq a(\alpha)/2,$$

where $a(\alpha) = 1 - P(0; \alpha, C, 1, 0)$, $C > 0$, $1 < \alpha < 2$, and the constant $N(\tau, 1)$ is defined by the lemma 2.5. Then from the formula (2.118) obtain

$$A = \sum_{k=1}^{K^*} P(M_{2^k} > n^{\delta-\gamma}2^{k-2}) + \sum_{k=K^*+1}^{\infty} P(M_{2^k} > n^{\delta-\gamma}2^{k-2}). \tag{2.119}$$

As

$$\sum_{k=1}^{K^*} P(M_{2^k} > n^{\delta-\gamma}2^{k-2}) \leq K^* P(M_{2^{K^*}} > n^{\delta-\gamma}/2), \tag{2.120}$$

so the formulas (2.119), (2.120) and the lemma 2.2 lead to

$$A \leq \frac{2K^*}{a(\alpha)} P\left(\sum_{k=1}^{2^{K^*+1}} \Delta x(k) > n^{\delta-\gamma}/2 \right) +$$

$$+ \frac{2}{a(\alpha)} \sum_{k=K^*+1}^{\infty} P\left(\sum_{j=1}^{2^{k+1}} \Delta x(j) > n^{\delta-\gamma}2^{k-2} \right) = A_1 + A_2. \tag{2.121}$$

From the lemma 2.5 obtain that for

$$\forall \tau, \ 1 < \tau < \alpha < 2, \ \exists N(\tau, 1) \in N = \{1, 2, ...\}, \ Q(\tau, 1)$$

so that for $\forall k, \ k \geq \log_2 N(\tau, 1) - 1,$

$$P\left(\sum_{j=1}^{2^{k+1}} \Delta x(j) > s\right) \leq \frac{2^{k+1} Q(\tau, 1)}{s^\tau} \tag{2.122}$$

for $s \geq 2^{(k+1)/\tau}((k+1)ln2)^2.$

In the case $A_1 \ \ s = n^{\delta-\gamma}/2.$ As $\delta > \gamma$ so $\exists N_1 : \ \forall n \geq N_1$

$$n^{\delta-\gamma}/2 \geq 2^{(K^*+1)/\tau}((K^* + 1)ln2)^2.$$

In the case $A_2 \ \ s = n^{\delta-\gamma}2^{k-2}.$ So for $\delta > \gamma$ obtain that $\exists N_2 : \forall n \geq N_2$ and for $\forall k \geq K^*$

$$n^{\delta-\gamma}2^{k-2} \geq 2^{(k+1)/\tau}((k+1)ln2)^2.$$

Then for $\forall n \geq \max(N_1, N_2)$ from (2.121), (2.122) obtain that

$$A_1 + A_2 \leq \frac{2}{a(\alpha)}\left(4K^*\frac{2^{K^*+1}Q(\tau, 1)}{(n^{\delta-\gamma})^\tau} + \sum_{k=K^*+1}^{\infty} \frac{2^{k+1}Q(\tau, 1)}{(n^{\delta-\gamma}2^{k-2})^\tau}\right) =$$

$$= \frac{2Q(\tau, 1)}{a(\alpha)(n^{\delta-\gamma})^\tau}\left(4K^*2^{K^*+1} + 2^{2\tau+1}\sum_{k=K^*+1}^{\infty} 2^{(1-\tau)k}\right). \tag{2.123}$$

Denoting

$$L(\alpha, \tau, 1, K^*) = \frac{2Q(\tau, 1)}{a(\alpha)}\left(4K^*2^{K^*+1} + 2^{2\tau+1}\sum_{k=K^*+1}^{\infty} 2^{(1-\tau)k}\right),$$

from the formulas (2.118), (2.121), (2.123) obtain that

$$P_1(n) \leq \frac{L(\alpha, \tau, 1, K^*)}{(n^{\delta-\gamma})^\tau}.$$

So the equality (2.117) true.

Suppose now that $\gamma > 1 - 1/\alpha$ or $\delta < \gamma.$ Divide the set $G = \{(\gamma, \delta) : 0 < \delta < \gamma$ or $\gamma > 1 - 1/\alpha\}$ into the nonintersecting subsets $G_{1,\alpha} = \{(\gamma, \delta) : \delta \geq 1 - 1/\alpha, \gamma > 1 - 1/\alpha\}$ and $G_{2,\alpha} = \{(\gamma, \delta) : 0 < \delta < \gamma, \delta < 1 - 1/\alpha\}, \ G_{1,\alpha} \cup G_{2,\alpha} = G.$

Consider the case when $(\gamma, \delta) \in G_{1,\alpha}$. From the formula (2.103) obtain that for $\forall M,\ M \in N = \{1, 2, ...\}$,

$$p_n = P\left(\sup_{m>0} \sum_{k=1}^{m}\left(n^{1-1/\alpha-\delta}\Delta x(k) + \sum_{j=1}^{n}\frac{\Delta x(k,j)}{n^{1/\alpha}} - n^{1-1/\alpha-\gamma}\right) > 0\right) \geq$$

$$\geq P\left(\max_{0<m\leq M} \sum_{k=1}^{m}\left(n^{1-1/\alpha-\delta}\Delta x(k) + \sum_{j=1}^{n}\frac{\Delta x(k,j)}{n^{1/\alpha}}\right) > Mn^{1-1/\alpha-\gamma}\right).$$

Analogously to the proof of the previous theorem in the case $(\gamma, \delta) \in G_1$ (with the single correction that $b=\alpha,\ 1<\alpha<2$, and $\lambda(k),\ k \geq 1$, are i.i.d.r.v.'s with the common stable d.f. $P(u; \alpha, C, 1, 0))$ obtain that for $\forall \epsilon > 0\ \exists N^*: \forall n \geq N^*$

$$p_n \geq 1 - 2\epsilon. \tag{2.124}$$

Suppose that $(\gamma, \delta) \in G_{2,\alpha}$. In this case for $\forall M,\ M \in N = \{1, 2, ...\}$ the following inequality

$$p_n = P\left(\sup_{m>0} \sum_{k=1}^{m}\left(\Delta x(k) + n^{\delta-(1-1/\alpha)}\sum_{j=1}^{n}\frac{\Delta x(k,j)}{n^{1/\alpha}} - n^{1-1/\alpha-\gamma}\right) > 0\right) \geq$$

$$\geq P\left(\max_{0<m\leq M} \sum_{k=1}^{m}\left(\Delta x(k) + n^{\delta-(1-1/\alpha)}\sum_{j=1}^{n}\frac{\Delta x(k,j)}{n^{1/\alpha}}\right) > Mn^{1-1/\alpha-\gamma}\right).$$

is true. Analogously to the proof of previous theorem in the case $(\gamma, \delta) \in G_2$ (with the corrections for b and $\lambda(k),\ k \geq 1$) obtain the inequality (2.124).

Chapter 3

New approaches to heavy tails asymptotic analysis

Modern research of the queueing systems and risk models show that the distribution tail $\overline{A}(x) = 1 - A(x)$ of the serving times or the risks of the insurance companies are "heavy" (that is they have not finite exponential moments: $\int_0^\infty \exp(at)dA(t) = \infty$, $a > 0$). This circumstance attracts the interest of manifold researchers to the asymptotic analysis of "heavy" tails in different queueing and risk models [6, 20, 39, 63]. A main part of the considered models is closely connected with the queueing systems $M|G|1|\infty$, $GI|GI|1|\infty$.

A transition to a more wide class of the models (and the cooperative effects in them) demand new research methods. One of these methods is the introduction of r.v.'s indexes. This method is analogous by a nature to the methods of the test functions and the probability metrics [99, 37]. A specific of listed methods is the following: these methods allow investigating properties of a stochastic model without a calculation of its distributions. In this chapter we analyze the queueing model $GI|GI|m|\infty$ and the risk model under the stochastic interest force by means of r.v.'s indexes. The indexes allow obtaining new results under more natural conditions.

At the end of this chapter, we construct a model of a multi-server queueing system with competition between servers. A comparison between main objective functions of the classical multi-server system and the system with the competition is made. For this purpose the asymptotic of the ability to handle for large number of the servers and for the tail of stationary distribution of waiting time are constructed.

3.1. Investigation of applied probability models by means of random variables indexes

Indexes of random variables and their properties

Define for r.v. X with d.f. F

1. $ind_*(X) = \sup\{r : EX^r < \infty\}$,

2. $ind^*(X) = \sup\{a : E\exp(X^a) < \infty\}$.

If r.v. X is distributed on $(-\infty, \infty)$ then put

$$ind_*(X) = ind_*(X^+), \quad ind_a(X) = ind_a(X^+),$$

where $X^+ = \max(0, X)$. So these indexes describe right tail properties of r.v. X.

List without proves some obvious properties, allowing to calculate r.v.'s indexes and to estimate distribution tails via them.

Lemma 3.1. *Suppose that X is r.v. with d.f. F, $\overline{F} = 1 - F$.*
1) If $\overline{F}(x) = L(x)x^{-\alpha}$ (for some $\alpha > 0$), where $L(x)$ is a slowly varying function then $ind_*(X) = \alpha$.
2) If $\overline{F}(x) = \exp(-c(x)x^a)$, where $c(x) \to c$, $c > 0$, for $x \to \infty$ then $ind^*(X) = a$.

Lemma 3.2. *Suppose that X is r.v. with d.f. F.*
1) If $0 < ind_*(X) = \alpha < \infty$ then for any positive number $b < \alpha$ there exists $B > 0$ so that for $x > 1$ the inequality $\overline{F}(x) \le Bx^{-b}$ is true.
2) If $0 < ind^*(X) = \alpha < \infty$ then for any positive number $b < \alpha$ there is $B > 0$ so that for $x > 1$ the inequality $\overline{F}(x) \le B\exp\{-x^b\}$ takes place.

In a formulation of the next lemma use the designation ind, which may be understood as ind_* so ind^*.

Lemma 3.3. *1) Suppose that X, Y are r.v.'s, satisfying the conditions*

$$0 < ind(X), \ ind(Y) < \infty, \quad P(X \ge 0, \ Y \ge 0) = 1, \tag{3.1}$$

then the equality $ind(X + Y) = \min(ind(X), ind(Y))$ is true.
2) Suppose that X, Y are independent r.v.'s, satisfying the conditions $0 < ind(X) <$

∞, $P(X \geq 0,\ Y \geq 0) = 1$. *then the equality* $ind(X - Y) = ind(X)$ *is true.*

3) Suppose that X, Y are independent r.v.'s, satisfying the conditions (3.1), then the inequalities

$$ind_*(\min(X,\ Y)) \geq ind_*(X) + ind_*(Y),$$

$$ind^*(\min(X,\ Y)) \geq \max(ind^*(X), ind^*(Y))$$

are true. If additionally we suggest that r.v.'s $X,\ Y$ satisfy the conditions of the statements 1) and 2) from the lemma 3.1 then these inequalities transform into the equalities.

4) Suppose that X, Y are r.v.'s, satisfying the condition (3.1), then the equality $ind(\max(X,\ Y)) = \min(ind(X),\ ind(Y))$ is true.

5) Suppose that X, Y are r.v.'s, satisfying the conditions (3.1), and for some $T \geq 0$: $P(X \geq t) \geq P(Y \geq t),\ t \geq T$, then $ind(X) \leq ind(Y)$.

Proof. The statement *1)* is based on the corollaries of the binomial theorem and the Ung inequalities: for the natural numbers $p,\ q$ and the positive numbers $x,\ y$

$$(x + y)^{p/q} \leq \left[x^{1/q} + y^{1/q}\right]^p \leq$$

$$\leq \sum_{i-0}^{p} C_p^i \left(\frac{i x^{p/q}}{p} + \frac{(p - i) y^{p/q}}{p}\right) \leq 2^p (x^{p/q} + y^{p/q}).$$

The statement *2)* is based on the inequality

$$Ef(X) \geq Ef\left[(X - Y)^+\right] \geq \sum_{k=0}^{\infty} P(k \leq Y < k + 1) Ef\left[(X - k - 1)^+\right]$$

with $f(u) = u^b,\ u > 0,\ b > 0,\ f(u) = \exp(cu^a),\ u > 0,\ a, c > 0$. It is clear that this inequality is true in the conditions of this statement. The statement *3)* is based on the statements *1)* and *2)* of the lemma 3.1, see also the lemma 1 from [16]. The statement *4)* is based on the obvious inequality

$$\max(Ef(X),\ Ef(Y)) \leq E \max(f(X),\ f(Y)) \leq Ef(X) + Ef(Y)$$

for f chosen in the statement *2)*. The statement *5)* is obvious.

Lemma 3.4. *Suppose that X, Y are independent r.v.'s, satisfying the condition (3.1), then the equalities*

$$ind_*(X\,Y) = \min(ind_*(X),\ ind_*(Y)),\ ind_*(cX) = ind_*(X),\ c > 0$$

are true.

Proof. This lemma statement is based on the equalities

$$E(XY)^a = EX^a\,EY^a,\ \ E(cX)^a = c^a\,EX^a,\ \ a, c > 0.$$

It is clear that the equalities are true in the lemma conditions.

Lemma 3.5. *1) Suppose that X, Y are independent r.v.'s, satisfying the condition (3.1), then for $c > 0$ the following statements*

$$ind^*(X^c) = \frac{ind^*(X)}{c},\ ind^*(cX) = ind^*(X),$$

$$\frac{1}{ind^*(X\,Y)} \le \frac{1}{ind^*(X)} + \frac{1}{ind^*(Y)}. \tag{3.2}$$

are true.

2) Suppose that r.v. Y_0 satisfies the conditions: $0 < ind^*(Y_0) < \infty$, $P(Y_0 > 0) = 1$ and the condition 2) of the lemma 3.1. Then the inequality

$$ind^*\left(\prod_{i=0}^{n-1} Y_i\right) \le \frac{ind^*(Y_0)}{n}$$

is true.

Proof. The equalities in the formulas (3.2) are obvious. Using the Ung inequality

$$ft \le \frac{f^s}{s} + \frac{t^r}{r},\ \ s > 1,\ t > 1,\ \frac{1}{s} + \frac{1}{r} = 1,\ \ f, t > 0,$$

and the statements *1)*, *5)* of the lemma 3.3 and the equalities in the formulas (3.2), obtain

$$ind^*(X_1 X_2) \ge \min\left(\frac{ind^*(X_1)}{s},\ \frac{ind^*(X_2)}{r}\right).$$

    Choose now

$$\frac{1}{s} = \frac{ind^*(X_2)}{ind^*(X_1) + ind^*(X_2)},\ \ \frac{1}{r} = \frac{ind^*(X_1)}{ind^*(X_1) + ind^*(X_2)}.$$

So we easily obtain the inequality in the formula (3.2). The statement *1)* is proved. The statement *2)* leads from the inequality

$$\prod_{i=0}^{n-1} Y_i \geq [\min(Y_0, \ldots, Y_{n-1})]^n,$$

the statement *3)* of the lemma 3.3 and from the formulas (3.2).

Multiserver queueing system

Consider the queueing system $GI|GI|m|\infty$ with m servers and recurrent input flow $t_0 = 0 \leq t_1 = t_0 + \xi_0 \leq t_2 = t_1 + \xi_1 \leq \ldots$ and serving times $\eta_0, \eta_1, \ldots$ Suppose that the random sequences $\{\xi_0, \xi_1, \ldots\}$, $\{\eta_0, \eta_1, \ldots\}$ are independent and consist of independent r.v.'s with following d.f.'s

$$P(\xi_0 < t) = F_1(t), \quad P(\eta_0 < t) = F_2(t) \tag{3.3}$$

and $F_1(0) = F_2(0)$.

Consider m-dimension Markov chain $W_n=(w_{n,1}, \ldots, w_{n,m})$, $n{\geq}0$, in which $w_{n,\,i}$ is the interval between the moment t_n and the moment when i servers become free of the 0-th, the 1-st, ..., the $(n-1)$-th input flow customers. This Markov chain was introduced by Kiefer and Wolfowitz [44] who proved the recurrent formula:

$$W_{n+1} = R((w_{n,\,1} + \eta_n - \xi_n)^+, \, (w_{n,\,2} - \xi_n)^+, \ldots, (w_{n,\,m} - \xi_n)^+), \quad n \geq 0,$$

$$W_0 = (0, \ldots, 0). \tag{3.4}$$

Here $R = R(v_1, \ldots, v_m)$ is the operator of the vector $(v_1, \ldots, v_m)$ components so that $v_1 \leq \ldots \leq v_m$, $a^+ = \max(0, \, a)$. Introduce auxiliary Markov chain $V_n = (v_{n,\,1}, \ldots, v_{n,\,m})$, $n \geq 0$, defined by the recurrent formula

$$V_{n+1} = R(v_{n,\,1} + \eta_n, \, v_{n,\,2}, \ldots, v_{n,\,m}), \quad n \geq 0, \quad V_0 = (0, \ldots, 0). \tag{3.5}$$

Using mathematical induction method it is easy to prove that

$$\left(v_{n,\,i} - \sum_{i=0}^{n-1} \xi_i\right)^+ \leq w_{n,\,i} \leq v_{n,\,i}, \quad 0 < n, \quad 1 \leq i \leq m, \tag{3.6}$$

and

$$V_m = (v_{m,\,1}, \ldots, v_{m,\,m}) = R(\eta_0, \ldots, \eta_{m-1}),$$

$$v_{m,\,i} = \max(\min(\eta_{j_1}, \ldots, \eta_{j_{m-i+1}}) : j_1, \ldots, j_{m-i+1} = \overline{0, m-1}),$$

$$1 \le i \le m. \tag{3.7}$$

From the formula (3.5) obtain that for $n \ge 0$

$$v_{n+1,\,i} = \max(v_{n,\,i},\ \min(v_{n,\,1} + \eta_n,\ v_{n,\,i+1})), \quad 1 \le i < m,$$

$$v_{n+1,\,m} = \max(v_{n,\,m},\ v_{n,\,1} + \eta_n). \tag{3.8}$$

Lemma 3.6. *If r.v. η_0 satisfies the condition of the statement 1) from the lemma 3.1 then*

$$ind_*(v_{m,\,i}) = (m - i + 1)\, ind_*(\eta_0), \quad 1 \le i \le m. \tag{3.9}$$

Proof. The lemma 3.6 statement is based on the formula (3.7), on the statement *1)* of the lemma 3.1 and on the statements *3)*, *4)* of the lemma 3.3.

Lemma 3.7. *If r.v. η_0 satisfies the condition of the statement 2) of the lemma 3.1 then*

$$ind^*(v_{m,\,i}) = ind^*(\eta_0), \quad 1 \le i \le m. \tag{3.10}$$

Proof. The statement of the lemma 3.7 is based on the formula (3.7), on the statement *2)* of the lemma 3.1 and on the statements *3)*, *4)* of the lemma 3.3.

Lemma 3.8. *If r.v. η_0 satisfies the condition 1) of the lemma 3.1 then*

$$ind_*(v_{n,\,i}) = (m - i + 1)\, ind_*(\eta_0), \quad 1 \le i \le m \le n. \tag{3.11}$$

Proof. It is clear that the formula (3.8) leads to the inequalities:

$$v_{n,\,i} \le v_{n+1,\,i} = v_{n,\,i} + \max(0,\ \min(v_{n,\,1} + \eta_n - v_{n,\,i},\ v_{n,\,i+1} - v_{n,\,i})) \le$$

$$\le v_{n,\,i} + \min(\eta_n,\ v_{n,\,i+1}), \quad 1 \le i < m, \tag{3.12}$$

$$v_{n,\,m} \le v_{n+1,\,m} \le v_{n,\,m} + \eta_n.$$

So from the formulas (3.12), from the lemma 3.6 and from the statements *1)*, *3)*, *5)* of the lemma 3.3 obtain

$$ind_*(v_{n,\,i}) \ge ind_*(v_{n+1,\,i}) \ge \min(ind_*(v_{n,\,i}),\ ind_*(\min(\eta_n,\ v_{n,\,i+1}))) \ge$$

$$\geq \min(ind_*(v_{n,\,i}),\ ind_*(\eta_n) + ind_*(v_{n,\,i+1})),$$

$$ind_*(v_{n,\,m}) \geq ind_*(v_{n+1,\,m}) \geq \min(ind_*(v_{n,\,m}),\ ind_*(\eta_n)). \tag{3.13}$$

Using the formulas (3.9), (3.13) and mathematical induction method, obtain the equalities (3.11).

Lemma 3.9. *If r.v. η_0 satisfies the condition of the statement 2) from the lemma 3.1 then*

$$ind^*(v_{n,\,i}) = ind^*(\eta_0),\ \ 1 \leq i \leq m \leq n. \tag{3.14}$$

Proof. The proof of this lemma repeats the proof of the lemma 3.8 practically the word by the word. Single difference is that besides of the lemma 3.6 we use here the lemma 3.7.

Using the inequality (3.6) and the statement *2)* of the lemma 3.3, obtain the following statements.

Theorem 3.1. *If r.v. η_0 satisfies the condition 1) of the lemma 3.1 then*

$$ind_*(w_{n,\,i}) = (m - i + 1)\,ind_*(\eta_0),\ \ 1 \leq i \leq m \leq n. \tag{3.15}$$

Theorem 3.2. *If r.v. η_0 satisfies the condition of the statement 2) from the lemma 3.1 then*

$$ind^*(w_{n,\,i}) - ind^*(\eta_0),\ \ 1 \leq i \leq m \leq n. \tag{3.16}$$

Remark 1. *The theorem 3.1 allows to clarify a nature of the statement 1.2 from [66] and the results of the first section on cooperative effects in multiserver queueing systems $M|M|n|\infty$ (considered like the aggregation of the one-server queueing systems $M|M|1|\infty$) [86].*

Risk model under stochastic interest force

Consider the random sequence

$$S_0 = x,\ \ S_{n+1} = \xi_n S_n + (1 - \eta_{n+1}),\ \ n \geq 0, \tag{3.17}$$

which describes the insurance company capital in the discrete moments $0, 1, \ldots$. Here x is the initial capital of the insurance company, $\eta_1, \eta_2, \ldots$ are its risks in the moments $1, 2, \ldots,$ $\xi_0,$ $\xi_1, \ldots$ are the inflation factors which transform the company capital from the moment 0 to the moment 1 and so on. Suppose that the company receives the unit sum of money during the time step. The sequences $\eta_1,$ $\eta_2, \ldots,$ $\xi_0,$ $\xi_1, \ldots$ are independent and consist of i.i.d.r.v.'s with common d.f., $\mathbf{P}(\xi_0 > 0) = 1.$

Denote $X_n = \eta_n - 1,$ $Y_n = \xi_{n-1}^{-1},$ $n \geq 1,$ then the model (3.17) may be rewritten in the form

$$S_0 = x, \quad S_n = Y_n^{-1} S_{n-1} - X_n, \quad n \geq 1. \tag{3.18}$$

Consider the function (the ruin probability in n steps)

$$\psi(x, n) = \mathbf{P}(\min_{0 \leq k \leq n} S_k < 0), \quad n \geq 0.$$

It is clear that the function $\psi(x, n)$ does not increase by $x \in [0, \infty)$ and does not decrease by $n \geq 0.$ By the definition $\psi(x, n) = P(U_n > x),$ where $U_0 = 0$ and

$$U_n = \max\left\{0, \max_{1 \leq k \leq n} \sum_{i=1}^{k} X_i \prod_{j=1}^{i} Y_j\right\}, \quad n \geq 1. \tag{3.19}$$

Define another Markov chain:

$$V_0 = 0, \quad V_n = Y_n \max\{0, \ X_n + V_{n-1}\}, \quad n \geq 1. \tag{3.20}$$

Lemma 3.10. *If for all $n \geq 0$ r.v.'s U_n and V_n coincide by the distribution:*

$$V_n \overset{d}{=} U_n, \quad n \geq 0, \tag{3.21}$$

then

$$\psi(x, n) = P(V_n > x), \quad n \geq 0. \tag{3.22}$$

Proof. For $n = 0$ the formula (3.21) is obvious. Suppose that (3.21) is true for $n = m - 1 \geq 0$ that is $V_{m-1} \overset{d}{=} U_{m-1}.$ Choose the equality (3.21) for $n = m.$ Remind that $\{X_n, n \geq 1\}$ and $\{Y_n, n \geq 1\}$ are the independent random sequences and each of them consists of i.i.d.r.v.'s. Accordingly to the definition (3.19) replace X_i and Y_j in U_{m+1} by X_{m+1-i} and Y_{m+1-j} correspondingly:

$$U_m = \max\left\{0, \max_{1 \leq k \leq m} \sum_{i=1}^{k} X_i \prod_{j=1}^{i} Y_j\right\} \overset{d}{=}$$

$$\overset{d}{=} \max\left\{0, \ \max_{1\le k\le m} \sum_{i=1}^{k} X_{m+1-i} \prod_{j=1}^{i} Y_{m+1-j}\right\} =$$

$$= \max\left\{0, \ \max_{1\le k\le m} \sum_{i^*=m+1-k}^{m} X_{i^*} \prod_{j^*=i^*}^{m} Y_{j^*}\right\} =$$

$$= Y_m \max\left\{0, X_m + \max_{1\le k\le m-1} \left\{0, \ \sum_{i^*=m+1-k}^{m-1} X_{i^*} \prod_{j^*=i^*}^{m-1} Y_{j^*}\right\}\right\} =$$

$$= Y_m \max\{0, \ X_m + V_{m-1}\}.$$

Consequently the formula (3.21) is true for $n = m$.

The lemma 3.10 generalizes Lindley chain, describing one-server queueing system $GI|GI|1|\infty$ [49] onto risk model under the stochastic interest force (3.17).

Theorem 3.3. *Suppose that X_0, Y_0 are independent r.v.'s, satisfying the condition (3.1), then for $m \ge 0$ the following equality*

$$ind_*(U_m) = ind_*(V_m) = \min(ind_*(X_0), \ ind_*(Y_0))$$

is true.

Proof. From the lemmas 3.4, 3.10 and the statements 1), 4) of the lemma 3.3 obtain

$$ind_*(U_1) = ind_*(V_1) = ind_*(X_0 \, Y_0) = \min(ind_*(X_0), \ ind_*(Y_0)),$$
$$ind_*(V_1+X_1) = \min(ind_*(V_1), ind_*(X_1)) = \min(ind_*(X_0), ind_*(Y_0)),$$
$$ind_*(U_2) = ind_*(V_2) = \min(ind_*(V_1+X_1), ind_*(Y_1)) = \min(ind_*(X_0), ind_*(Y_0)), \dots$$

Using the mathematical induction method, it is easy to spread the the theorem 3.3 onto arbitrary $m > 0$.

Theorem 3.4. *Suppose that for some positive number ε*

$$P(X_0 > \varepsilon) = 1, \quad P(Y_0 > \varepsilon) = 1,$$
$$0 < ind^*(X_0) < \infty, \quad 0 < ind^*(Y_0) < \infty \tag{3.23}$$

and r.v. Y_0 satisfies the condition 2) of the lemma 3.1. Then for $m > 0$

$$f_1(m) \leq ind^*(U_m) \leq f_2(m),$$

$$f_1(m) = \frac{ind^*(X_0)ind^*(Y_0)}{ind^*(Y_0) + m\, ind^*(X_0)}, \quad f_2(m) = \frac{ind^*(Y_0)}{m},$$

$$f_1(m) \sim f_2(m), \quad m \to \infty.$$

Proof. Using the formula (3.19) and mathematical induction method, it is easy to prove that

$$U_m^{(1)} \leq U_m \leq U_m^{(2)}, \quad U_m^{(1)} = \varepsilon \prod_{i=1}^{m} Y_i, \quad U_m^{(2)} = \varepsilon^{-m} \sum_{k=1}^{m} X_k \varepsilon^k \prod_{i=1}^{m} Y_i$$

consequently

$$ind^*(U_m^{(2)}) \leq ind^*(U_m) \leq ind^*(U_m^{(1)}).$$

From the lemma 3.5 obtain the inequality

$$ind^*(U_m^{(1)}) \leq \frac{ind_*(Y_0)}{m} = f_2(m).$$

The statements in the lemmas *1)* 3.3, 3.5 lead to the formula

$$f_1(m) = \frac{ind^*(X_0)ind^*(Y_0)}{ind^*(Y_0) + m\, ind^*(X_0)} \leq ind^*(U_m^{(2)}).$$

Remark 2. *In the conditions of the theorem 3.3 the ruin probability asymptotic is defined by more heavier among r.v.'s X_0, Y_0 distributions tails. In the conditions of the theorem 3.4 the ruin probability asymptotic is defined by r.v. Y_0 distribution tail. The theorems 3.3, 3.4 give new results in the risk model under stochastic interest force in a comparison with [38, 51, 55].*

3.2. Multi-server queueing system with competition between servers

The queueing systems with the competition between servers or the customers are widely used in modern data transmission and mobile telephone networks [2]. But analytical investigation of the competition influence on the queueing systems characteristics was not made. In this subsection mathematical model of multiserver queueing system with the competition between servers is constructed. This system is compared with the classical multiserver queueing system in terms of their abilities to handle and the distribution tails of the stationary waiting times.

Consider multiserver queueing system $GI|GI|m|\infty$ with m servers. input flow $0 = t_1 \leq t_2 = t_1 + \xi_1 \leq t_3 = t_2 + \xi_2 \leq \ldots$ and serving times $\eta_1, \eta_2, \ldots$. The random sequences $\{\eta_1, \eta_2, \ldots\}$, $\{\xi_1, \xi_2, \ldots\}$ are independent and consist of independent r.v.'s with d.f.'s $G(t)$, $H(t)$ concentrated on $[0, \infty)$ and with the tails

$$P(\eta_k > t)=1 - G(t)=\overline{G}(t), P(\xi_k > t)=1 - H(t)=\overline{H}(t), \ k \geq 0. \tag{3.24}$$

Denote this system by A_m and suppose that it is empty at the moment t_1 of the first customer arrival.

On a base of the system A_m define the multiserver queueing system with the competition between the servers B_m as follows. Suppose that the first customer arrives in empty system B_m at the moment t_1 and asks all m servers how long do they would serve it. The customer receives the information on its possible serving times $\eta_1^{(1)}, \ldots, \eta_1^{(m)}$ at the servers. Then it chooses the server with the minimal serving time $\zeta_1 = \min(\eta_1^{(1)}, \ldots, \eta_1^{(m)})$. During the first customer serving all other servers do not work. The second customer arrives into the system B_m at the moment t_2 and receives the information on its possible serving times $\eta_2^{(1)}, \ldots, \eta_2^{(m)}$ at the servers and choose the server with the minimal serving time $\zeta_2 = \min(\eta_2^{(1)}, \ldots, \eta_2^{(m)})$. During its service all other servers do not work and so on. R.v.'s $\eta_i^{(j)}$, $j = 1, \ldots, m$, $i \geq 1$, are independent and have common d.f. $G(t)$. So the system B_m may be considered as the oneserver queueing system $GI|GI|1|\infty$ with the recurrent input flow $0 = t_1 \leq t_2 = t_1 + \xi_1 \leq t_3 = t_2 + \xi_2 \leq \ldots$ and with i.i.d. serving times $\zeta_1, \zeta_2, \ldots$,

$$P(\zeta_1 > x) = \overline{G}^m(x), \quad x \geq 0. \tag{3.25}$$

If input flow $0 = t_1 \leq t_2 = t_1 + \xi_1 \leq t_3 = t_2 + \xi_2 \leq \ldots$ is Poisson then denote A_m, B_m by A_m^*, B_m^* correspondingly.

Suppose that w_n^A, w_n^B, $n \geq 0$, are the waiting times of the $n-$th customer in the systems A_m, B_m correspondingly. Denote by

$$\overline{W}_m^A(x) = \lim_{n \to \infty} P(w_n^A > x), \ \overline{W}_m^B(x) = \lim_{n \to \infty} P(w_n^B > x)$$

the tails limit distributions of the waiting times in the systems A_m, B_m. Fix d.f. $G(t)$ and designate by

$$\lambda_m^A = \sup\{1/M\xi_1 : \lim_{x\to\infty} \overline{W}_m^A(x) = 0\},$$

$$\lambda_m^B = \sup\{1/M\xi_1 : \lim_{x\to\infty} \overline{W}_m^B(x) = 0\}$$

the maximal abilities to handle of the systems A_m, B_m.

Our problem is to make the asymptotic analysis for $m \to \infty$ of

$$K_m = \frac{\lambda_m^A}{\lambda_m^B}$$

and to analyze the asymptotic of the function $\overline{W}_m^B(x)$, $x \to \infty$, when d.f. $G(x)$ is subexponential. For the systems A_m^*, B_m^* with $\overline{G}(x) \in \mathcal{R}(-a)$, $a > 1$ (here $\mathcal{R}(-a)$ is the class of the regular varying functions with the index a defined below) we shall compare he asymptotic behavior of the functions $\overline{W}_m^B(x)$, $\overline{W}_m^A(x)$ for $x \to \infty$.

Preliminary information

List the class of d.f. which will be used in this subsection. Denote by $\mathcal{S}$ subexponential class of d.f. F defined on $(0, \infty)$, $\overline{F} = 1 - F$ so that

$$\int_0^x \overline{F}(t-u)dF(u) \sim 2\overline{F}(x), \quad x \to \infty. \tag{3.26}$$

Suppose that $\mathcal{L}$ is the class of differentiated on $[0, \infty)$ and slowly varying functions, denote

$$\mathcal{L}_1 = \{l(x) \in \mathcal{L} : \limsup_{x\to\infty} \frac{xl'(x)}{l(x)} < \infty\},$$

$$\mathcal{R}(a) = \{x^a l(x), \ l(x) \in \mathcal{L}\}, \quad -\infty < a < \infty.$$

Designate by $\mathcal{L}_*$ the class of d.f. F so that for any $M > 0$

$$\lim_{t\to\infty} \frac{\overline{F}(t+M)}{\overline{F}(t)} = 1. \tag{3.27}$$

By the definition [31] d.f. F defined on $(0, \infty)$ belongs to the class S_* if $F(x) < 1$, its mean $\mu = \int_0^\infty \overline{F}(t)dt < \infty$ and

$$\int_0^x \overline{F}(t-u)\overline{F}(u)du \sim 2\mu\overline{F}(x), \quad x \to \infty. \tag{3.28}$$

Denote

$$\overline{F}_I(x) = \frac{1}{\mu} \int_x^\infty \overline{F}(y)dy. \tag{3.29}$$

Accordingly to [76], [19], [31] list now main properties of the introduced classes (Karamata theorem, the lemmas 3.11, 3.12, 3.13 and the theorem 3.5).

The formula $\mathcal{S}_* \subset \mathcal{S} \subset \mathcal{L}_*$ is true and for $\overline{F} \in \mathcal{R}(a)$, $a < 0$, there is the inclusion $F \in \mathcal{S}$ and the Karamata theorem takes place: "if $l(x) \in \mathcal{L}$, then for $x_0 > 0$, $a > -1$

$$\int_{x_0}^x t^a l(t)dt \sim (a+1)^{-1}x^{a+1}l(x), \quad x \to \infty, \tag{3.30}$$

and for $a < -1$

$$\int_x^\infty t^a l(t)dt \sim -(a+1)^{-1}x^{a+1}l(x), \quad x \to \infty". \tag{3.31}$$

Lemma 3.11. *If d.f.* $F \in \mathcal{S}_*$ *then* $F \in \mathcal{S}$, $F_I \in \mathcal{S}$. *If d.f.* $F \in \mathcal{S}_*$ *and d.f.* F_+ *satisfies the condition* $\overline{F}(x) \sim \overline{F}_+(x)$, $x \to \infty$, *then* $F_+ \in \mathcal{S}_*$.

Denote $Q(x) = -\ln\overline{G}(x)$ and its derivative $Q'(x) = q(x)$. Say that $q(x)$ decreases conditionally to 0 if for some positive number X the function $q(x)$ monotonically decreases to zero, $x \geq X, x \to \infty$.

Lemma 3.12. *Each of the listed conditions is sufficient for the inclusion* $G \in \mathcal{S}_*$.
(a) $\limsup_{x\to\infty} xq(x) < \infty$.
(b) There exists $\delta \in (0,1)$ *and* $v \geq 1$ *so that* $Q(xy) \leq y^\delta Q(x)$ *for all* $x \geq v$, $y \geq 1$ *and* $\liminf_{x\to\infty} xq(x) \geq (2 - 2^\delta)^{-1}$.
(c) $q(x)$ *conditionally decreases to 0 and*

$$\lim_{x\to\infty} \int_0^x \exp(yq(x))\overline{G}(y)dy = \mu.$$

Corollary 3.1. *Suppose that*

$$\lim_{x\to\infty} q(x) = 0, \quad \lim_{x\to\infty} xq(x) = \infty.$$

If additionally one of the following conditions is true then $G \in \mathcal{S}_*$.
(a) $\limsup_{x\to\infty} xq(x)/Q(x) < 1$.
(b) $q(x) \in \mathcal{R}(-\delta)$, $\delta \in (0,1]$.
(c) $q(x)$ *conditionally decreases to 0 and* $Q(x) \in \mathcal{R}(\delta)$, $\delta \in (0,1)$.
(d) $q(x)$ *conditionally decreases to 0 and* $q(x) \in \mathcal{R}(0)$ *and* $Q(x) - xq(x) \in \mathcal{R}(1)$.

Lemma 3.13. *Suppose that X_1, X_2 are independent and nonnegative r.v.'s and $\overline{V}_i(x) = P(X_i > x)$, $i = 1, 2$, $V_1(x) \in \mathcal{L}_*$ then $P(X_1 - X_2 > x) = \overline{V}(x) \sim \overline{V}_1(x)$, $x \to \infty$.*

Consider now the system $GI|GI|1|\infty$ in which serving times η_k, $k \geq 0$, $P(\eta_k < x) = G(x)$ and interarrival intervals ξ_k, $k \geq 0$, $P(\xi_k > x) = H(x)$ create the independent sequences of i.i.d.r.v.'s with their own d.f.'s. Suppose that i.i.d.r.v.'s $X_k = \eta_k - \xi_k$, $k \geq 0$, have common d.f. $V(x)$, $-\infty < x < \infty$, with the mean m, $-\infty < m < 0$. Denote

$$S_n = X_1 + \ldots + X_n, \quad M = \max_{n \geq 0} S_n, \quad \overline{W}(x) = P(M > x).$$

Then $\overline{W}(x)$ is the tail of stationary distribution of the waiting time in the system $GI|GI|1|\infty$ and $\lim_{x \to \infty} \overline{W}(x) = 0$ if and only if

$$M\eta_1 < M\xi_1. \tag{3.32}$$

From this condition and the inequality (3.25) obtain

$$\lambda_m^B = \frac{1}{\int_0^\infty \overline{G}^m(t)dt}. \tag{3.33}$$

Suppose that d.f. $V(x) \in \mathcal{S}_*$. It is well known that in this case (see [6], [19], [21] and the references in these papers) that

$$\overline{W}(x) \sim \frac{1}{|m|} \int_x^\infty \overline{V}(t)dt, \quad x \to \infty. \tag{3.34}$$

Consider the system $M|GI|1|\infty$ with Poisson input flow (with the intensity λ). If $G_I \in \mathcal{S}$ (for G_I see the formula (3.29)) then the Embrechts-Veraverbeke formula [21] is true:

$$\overline{W}(x) \sim \frac{\lambda M\eta_1}{1 - \lambda M\eta_1} \overline{G}_I(x), \quad x \to \infty.$$

Consider now the Kiefer-Wolfowitz chain [44] $(w_{n,1}, w_{n,2}, \ldots, w_{n,m})$, $n \geq 0$, describing multiserver queueing system A_m. Here $w_{n,i}$ is the interval between the moment t_n and the moment when i servers become free of the 1-st,...,$(n-1)$-th customers of input flow. Then recurrent formula

$$(w_{n+1,1}, \ldots, w_{n+1,m}) = R((w_{n,1} + \eta_n - \xi_n)^+, (w_{n,2} - \xi_n)^+, \ldots,$$

$$(w_{n,\,m} - \xi_n)^+), \quad n \geq 0, \tag{3.35}$$

with $R = R(v_1, \ldots, v_m)$ - ordering operator of the vector $(v_1, \ldots, v_m)$ components, $v_1 \leq \ldots \leq v_m$. $a^+ = \max(0, a)$ is true. Necessary and sufficient condition of the Kiefer-Wolfowitz chain ergodicity (and so the existence of the limit distribution $\lim_{n \to \infty} P(w_{n,\,1} > t))$ is the inequality

$$M\eta_1 < mM\xi_1. \tag{3.36}$$

So the formula

$$\lambda_m^A = \frac{m}{\int_0^\infty \overline{G}(t)dt} \tag{3.37}$$

takes place and accordingly to the equalities (3.33), (3.37) obtain

$$K_m = \frac{m \int_0^\infty \overline{G}^m(t)dt}{\int_0^\infty \overline{G}(t)dt}. \tag{3.38}$$

In [96, theorem 6] it is proved that

Theorem 3.5. *Suppose that in the system A_m^* for some $a > 1$ d.f. $G(x) \in \mathcal{R}(-a)$ and the ergodicity condition (3.36) is true then the function $\overline{W}_m^A(x) \in \mathcal{R}(-ma+m)$.*

Main results and proves

Theorem 3.6. *Suppose that for some $a > 0$, $b > 0$:*

$$G(x) \sim ax^b, \quad x \to 0 \tag{3.39}$$

then in the system B_m for $m \to \infty$ the following formula is true:

$$\text{if } 0 < b \leq 1, \text{ then } K_m = O(m^{1-1/b}), \tag{3.40}$$

$$\text{if } b \geq 1, \text{ then } 1/K_m = O(1/m^{1-1/b}). \tag{3.41}$$

Proof. Suppose that $0 < b \leq 1$, $a > 0$ are fixed. Then there exist the positive numbers α, m_0 so that for $0 < t < 1/m_0$ the inequality $G(t) \geq \alpha t^b$ is true and so $\overline{G}(t) \leq \exp(-\alpha t^b)$. Suppose that $m > m_0$ and denote

$$J_m = \int_0^\infty m\overline{G}^m(t)dt = U_m + V_m, \quad U_m = \int_0^{1/m} m\overline{G}^m(t)dt,$$

$$V_m = \int_{1/m}^{\infty} m\overline{G}^m(t)dt.$$

Estimate the quantity U_m for $m \to \infty$:

$$U_m \le \int_0^{1/m} m\exp(-\alpha mt^b)dt = \int_0^{1/m} m\exp(-((\alpha m)^{1/b}t)^b)\frac{dt(\alpha m)^{1/b}}{(\alpha m)^{1/b}} =$$

$$= \frac{m}{(\alpha m)^{1/b}}\int_0^{(\alpha m)^{1/b}/m}\exp(-v^b)dv \sim \frac{m}{(\alpha m)^{1/b}}\int_0^{\infty}\exp(-v^b)dv = O(m^{1-1/b}).$$

Estimate now the quantity V_m for $m \to \infty$:

$$V_m \le m\overline{G}^{m-1}(1/m)\int_{1/m}^{\infty}\overline{G}(t)dt \le$$

$$\le m\exp(-\alpha(m-1)m^{-b})\int_0^{\infty}\overline{G}(t)dt = o(m^{1-1/b}).$$

Then $J_m = O(m^{1-1/b})$ for $m \to \infty$ and

$$K_m = J_m/J_1 = O(m^{1-1/b}), \quad m \to \infty.$$

The formula (3.40) is proved.

Suppose that $a > 0$, $b \ge 1$ are fixed. Then there exist the positive numbers α, m_0 so that for $0 < t < 1/m_0$ the inequality $G(t) \le \alpha t^b$ is true and so $\overline{G}(t) \ge 1 - \alpha t^b$. Put $m > m_0$ and obtain

$$J_m \ge \int_0^{1/m^{1/b}} m(1-\alpha/m)^m dt \sim m^{1-1/b}\exp(-\alpha), \ m \to \infty$$

and

$$1/K_m = J_1/J_m = O(1/m^{1-1/b}), \quad m \to \infty.$$

The formula (3.41) is proved.

Consider now the asymptotic behavior of the function $\overline{W}_m^B(x)$ for $x \to \infty$ in the system B_M^*.

Theorem 3.7. *Suppose that in the system B_m^* d.f. $G(x) \in \mathcal{R}(-a)$, $a > 1$ and the ergodicity condition $\lambda b_m < 1$ with $b_m = \int_0^{\infty}\overline{G}^m(x)dx$ is true. Then*

$$\overline{W}_m^B(x) \sim \frac{\lambda x \overline{G}^m(x)}{(1-\lambda b_m)(ma-1)}, \quad x \to \infty \tag{3.42}$$

and consequently the function $\overline{W}_m^B(x) \in \mathcal{R}(-ma+1).$

Proof. As d.f. $G(t) \in \mathcal{R}(-a)$, $a > 1$, so for some function $l(t) \in \mathcal{L}$ (and consequently $l^m(t) \in \mathcal{L}$) $\overline{G}^m(t) = l^m(t)t^{-ma}$. Using the Karamata theorem for $x \to \infty$ obtain

$$\frac{1}{b_m} \int_x^\infty \overline{G}^m(t)dt \sim \frac{l^m(x)x^{-ma+1}}{b_m(ma-1)} = \frac{x\overline{G}^m(x)}{b_m(ma-1)}. \tag{3.43}$$

Then d.f.

$$1 - \frac{\int_x^\infty \overline{G}^m(t)dt}{b_m}$$

is subexponential and from the ergodicity condition $\lambda b_m < 1$ we obtain the Embrechts-Veraberbeke formula

$$\overline{W}_m^B(x) \sim \frac{\lambda b_m \int_x^\infty \overline{G}^m(t)dt}{b_m(1 - \lambda b_m)}, \quad x \to \infty. \tag{3.44}$$

Put the formula (3.43) into (3.44) and obtain the formula (3.42).

Corollary 3.2. *The theorem 3.6 shows how the ratio between the abilities to handle in the systems A_m (without the competition) and B_m (with the competition) depends on the parameter b.*

Consider the following examples of d.f. $G(x)$, $x \geq 0$, satisfying the condition (3.39): Weibull distribution

$$\overline{G}(x) = \exp(-ax^b), \quad 0 < b < 1,$$

and Burr distribution

$$\overline{G}(x) = (1 + cx^b)^{-a/c}, \quad 0 < a,\ 0 < b \leq 1,\ 0 < c < ab.$$

Remark 3. *The theorems 3.5, 3.7 comparison shows that in the system B_m^* with the competition the tail of the stationary waiting time distribution is lighter then in the system A_m^* without the competition.*

Theorem 3.8. *Suppose that in the system B_m the ergodicity condition $b_m < M\xi_1$ is true and the functions $Q(t)$, $q(t)$ satisfy one of the lemma 3.12 conditions (a), (b) or one of the corollary 3.1 conditions (a) − (d) then*

$$\overline{W}_m^B(x) \sim \frac{\int_x^\infty \overline{G}^m(t)dt}{M\xi_1 - b_m}, \quad x \to \infty. \tag{3.45}$$

Proof. Suppose that the functions $Q(t)$. $q(t)$ satisfy one of the conditions (a), (b) from the lemma 3.12 or one of the conditions $(a) - (d)$ of the corollary 3.1. Direct calculations show that the functions $mQ(t)$, $mq(t)$ also satisfy one of these conditions. So in the suggestions of the theorem 3.8 d.f. $P(\zeta_1 \leq t) \in S_*$ from the conclusion $S_* \subset S \subset \mathcal{L}_*$ and from the lemmas 3.11, 3.13 obtain

$$P(\zeta_1 - \xi_1 \leq t) \in S_*, \quad P(\zeta_1 - \xi_1 > t) \sim P(\zeta_1 > t), \quad t \to \infty.$$

So the ergodicity condition $b_m < M\xi_1$ and the formula (3.34) lead to

$$\overline{W}^B_m(x) \sim \frac{\int_x^\infty P(\zeta_1 - \xi_1 > t)dt}{M\xi_1 - b_m} \sim \frac{\int_x^\infty P(\zeta_1 > t)dt}{M\xi_1 - b_m}, \quad x \to \infty.$$

Corollary 3.3. *Suppose that in the system B_m the ergodicity condition $b_m < M\xi_1$ is true and the function $q(t)$ satisfies the following conditions: $q(t)$ is continuously differentiated and for some $\delta \in (0,1)$ the inclusion $q(t) \in \mathcal{R}(-\delta)$ is true and the derivative $q'(t) = o(q^2(t))$, $t \to \infty$. Then*

$$\overline{W}^B_m(x) \sim \frac{\overline{G}^m(x)}{(M\xi_1 - b_m)mq(x)}, \quad x \to \infty. \tag{3.46}$$

Proof. Directly check that d.f. $G(t)$ satisfies the condition (b) of the corollary 3.1. So the theorem 3.8 leads to the formula (3.45). From the Karamata theorem and the inequality $0 < \delta < 1$ obtain the inclusion $Q(t) \in \mathcal{R}(1 - \delta)$. Using this inclusion and the condition $q'(t) = o(q^2(t))$, $t \to \infty$, calculate

$$I_m(x) = \int_x^\infty \exp(-mQ(t))dt = \int_x^\infty \exp(-mQ(t))\frac{mq(t)}{mq(t)}dt =$$

$$= -\int_x^\infty d\exp(-mQ(t))\frac{1}{mq(t)} =$$

$$= \frac{\exp(-mQ(x))}{mq(t)} - \int_x^\infty \exp(-mQ(t))\frac{q'(t)}{q^2(t)}dt =$$

$$= \frac{\exp(-mQ(x))}{mq(x)} - o(I_m(x)), \quad x \to \infty.$$

So we obtain

$$\int_x^\infty \overline{G}^m(t)dt = I_m(x) \sim \frac{\exp(-mQ(x))}{mq(x)}, \quad x \to \infty.$$

Put this formula into (3.45) and obtain the formula (3.46).

Corollary 3.4. *Suppose that in the system B_m the ergodicity condition $b_m < M\xi_1$ is true and for some $l(x) \in \mathcal{L}_1, a > 1$ the equality $\overline{G}(x) = l(x)x^{-a}$ takes place then*

$$\overline{W}^B_m(x) \sim \frac{x\overline{G}^m(x)}{(M\xi_1 - b_m)(ma - 1)}, \quad x \to \infty. \tag{3.47}$$

Proof. From the condition of this corollary obtain that

$$Q(x) = a\ln x - \ln l(x), \quad q(x) = \frac{a}{x} - \frac{l'(x)}{l(x)} \sim \frac{a}{x}, \quad x \to \infty.$$

So the condition (a) of the lemma 3.12 is true and then d.f. $G(t) \in \mathcal{S}_*$. The ergodicity condition $b_m < M\xi_1$ and the formula (3.34) leads to (3.45). So to end the formula (3.47) proof it is enough to use the Karamata theorem.

Remark 4. *The theorems 3.6 - 3.8 show that the comparison of the systems B_m and A_m by their abilities to handle is defined by the function $G(x)$ behavior for $x \to 0$. And the comparison of the tails of the stationary waiting time distributions is defined by the function $G(x)$ behavior for $x \to \infty$.*

Chapter 4

Cooperative effects in reliability models

In this chapter, the counteraction effects in reliability systems are considered. At the beginning of the chapter, we analyze the phase transition phenomenon in aggregated renewal systems. These systems may be considered as systems of mutual reserve. In some regions of the parameter set, the aggregated systems begin to insure each other. This effect disappears out of this parametric region.

In a calculation of the static network reliability (considered later in this chapter) it is necessary to use about 2^n arithmetical operations. Here n is a number of the elements in the network. This calculation difficulty may be overcome if we replace the reliability by a necessary reserve volume with a simple upper bound $c \ln n$, $c > 0$. This fact may be obtained by sufficiently simple analytical calculations and estimates.

At the end of this chapter, some examples of a concrete technical reliability problem solution are considered. This example was realized when the standard optimization methods did not give positive results only because of their large complexity. At the same time, the application of the cooperative effects allowed obtaining sufficiently simple and reasonable solutions practically without complex calculations. This example confirms a fact which is intuitively clear to designers: in the control of complex systems reliability with continuously varying parameters, it is necessary to pay serious attention to the choice of their structure.

4.1. Renewal systems with common reserve

In [81, 84, 75] the fact of the strong counteraction between aggregated subsystems is recognized and the quantitative estimates of this effect were obtained for different queueing and insurance and reserve systems. In this subsection the phase transition connected with this counteraction in the aggregated renewal reserve system is considered. Some numerical results, describing this phenomenon, are analyzed. Obtained results may be used in telecommunication and computer networks and an another technical systems.

Consider the system of the unloaded duplication and the restoration. This system has one working place with he break intensity λ, one repair place with the renewal intensity μ, $\rho = \lambda/\mu$, and two elements. Take n independent copies of this system and aggregate them into the system of the unloaded duplication and the restoration with n working places, n repair places and $2n$ elements. Suppose that $k = k(t)$ is the current number of elements in the repair phase of the aggregated system. The process $k(t)$ is the birth and death process with the state set $\{0, 1, \ldots, 2n\}$ and the following birth λ_k and death μ_k intensities:

$$\lambda_k = \begin{cases} n\lambda, & 0 \leq k \leq n, \\ (2n - k)\lambda, & n < k \leq 2n, \end{cases} \tag{4.1}$$

$$\mu_k = \begin{cases} k\mu, & 0 \leq k \leq n, \\ n\mu, & n < k \leq 2n. \end{cases} \tag{4.2}$$

Investigate the behavior of the probability $P_n = P_n(\rho)$ of elements presence at all working places for different $\rho = \rho(n)$ and $n \to \infty$. P_n may be interpreted as the probability of the work at all n aggregated subsystems. The main properties of P_n are given in the following statements.

Theorem 4.1. *If $\rho = const$ then*

$$\lim_{n \to \infty} P_n = \frac{1}{2}, \quad \rho = 1, \quad \lim_{n \to \infty} P_n = 1, \quad \rho < 1, \quad \lim_{n \to \infty} P_n = 0, \quad \rho > 1.$$

Proof. From the formulas (4.1), (4.2) obtain that the stationary probabilities π_k, $0 \leq k \leq 2n$, of the process $k(t)$ satisfy the formulas:

$$n\mu\pi_{n+1} = n\lambda\pi_n, \quad n\lambda\pi_{n-1} = n\mu\pi_n$$

and so

$$\pi_{n+1} = \pi_n \rho^1, \qquad \pi_{n-1} = \pi_n \rho^{-1}. \tag{4.3}$$

Analogously obtain

$$(n-1)\lambda\pi_{n+1} = n\mu\pi_{n+2}, \quad n\lambda\pi_{n-2} = (n-1)\mu\pi_{n-1},$$

and so

$$\pi_{n+2} = \pi_{n+1}\left(\frac{n-1}{n}\right)\rho, \quad \pi_{n-2} = \pi_{n-1}\left(\frac{n-1}{n}\right)\rho^{-1},$$

consequently

$$\pi_{n+2} = \left(1-\frac{1}{n}\right)\rho^2\pi_n, \quad \pi_{n-2} = \left(1-\frac{1}{n}\right)\rho^{-2}\pi_n. \tag{4.4}$$

From the formulas (4.3), (4.4), continuing by induction, obtain

$$\pi_{n+k} = \pi_n\rho^k a_k, \quad \pi_{n-k} = \pi_n\rho^{-k} a_k,$$

$$a_k = \prod_{j=0}^{k-1}\left(1-\frac{j}{n}\right), \quad 1 \le k \le n. \tag{4.5}$$

Denote $F_n(\rho) = \sum_{k=1}^{n} a_k\rho^k$. From the formula (4.5) and the equality $P_n = \pi_0 + \pi_1 + ... + \pi_n$ obtain

$$P_n - P_n(\mu) = \frac{1 + F_n(\rho^{-1})}{1 + F_n(\rho) + F_n(\rho^{-1})}. \tag{4.6}$$

Investigate the function $F_n(\rho)$ for $\rho = 1$:

$$F_n(1) \ge \sum_{1 \le k \le n/2}\prod_{j=0}^{k-1}\left(1-\frac{j}{n}\right) \ge \sum_{1 \le k \le n/2}\exp\left(-\sum_{j=0}^{k-1}\frac{2j\ln 2}{n}\right) \ge$$

$$\ge \sum_{2 \le k \le n/2+1}\exp\left(-\frac{k^2\ln 2}{n}\right) \ge J_n,$$

$$J_n = \sqrt{n}\int_{\sqrt{2/n}}^{\sqrt{n/2}+\sqrt{1/n}}\exp(-v^2\ln 2)dv \to \infty, \quad n \to \infty.$$

So $F_n(1) \to \infty$, $n \to \infty$, and from the formula (4.6) obtain

$$P_n(1) \to \frac{1}{2}, \quad n \to \infty. \tag{4.7}$$

Now consider the cases $\rho > 1$ and $\rho < 1$. Investigate the function $F_n(\rho)$ for $\rho > 1$, choosing b from the conditions: $0 < b < 1$, $\rho(1 - b) > 1$, then

$$F_n(\rho) \geq \sum_{1 \leq k \leq bn} (\rho - \rho b)^k \geq (\rho - \rho b)^{[bn]} \to \infty, \quad n \to \infty. \tag{4.8}$$

Here $[a]$ is the integer part of the real number a. The obvious inequality takes place:

$$F_n(\rho^{-1}) \leq \sum_{1 \leq k \leq n} \rho^{-k} \leq \frac{\rho}{(\rho - 1)} < \infty. \tag{4.9}$$

From the formulas (4.6), (4.8), (4.9) obtain

$$P_n(\rho) \to 0, \quad n \to \infty. \tag{4.10}$$

Using the formulas (4.6), (4.8), (4.9) and the replacement $\rho \to \rho^{-1}$ it is easy to obtain that for $\rho < 1$

$$P_n(\rho) \to 1, \quad n \to \infty. \tag{4.11}$$

The formulas (4.7), (4.10), (4.11) lead to the theorem 4.1 statement.

Theorem 4.2. *If $\rho = \exp(-\alpha_n)$, $\alpha_n = n^{-1/2+\varepsilon}$, $1/2 > \varepsilon > 0$ then*

$$\lim_{n \to \infty} P_n = 1.$$

If $\rho = \exp(-\alpha_n)$, $\alpha_n = n^{-1/2-\varepsilon}$, $\varepsilon > 0$ then

$$\lim_{n \to \infty} P_n = \frac{1}{2}.$$

Proof. Define the auxiliary function

$$R_n(\rho) = \sum_{1 \leq k \leq n/2} a_k \rho^k, \quad T_n(\rho) = \sum_{n/2 < k \leq n} a_k \rho^k, \quad F_n(\rho) = R_n(\rho) + T_n(\rho).$$

It is clear that

$$1 - t \leq e^{-t}, \, t \geq 0, \quad 1 - t \geq e^{-ct}, \, 0 \leq t \leq \frac{1}{2}, \quad c = 2\ln 2 > 1. \tag{4.12}$$

From the formula (4.12) obtain

$$\exp\left(-c\sum_{j=0}^{k-1} \frac{j}{n}\right) \leq a_k, \, 1 \leq k \leq n/2, \quad \exp\left(-\sum_{j=0}^{k-1} \frac{j}{n}\right) \geq a_k, \, k \geq 1,$$

consequently

$$\exp\left(-\frac{ck^2}{2n}\right) \le a_k, \ 1 \le k \le \frac{n}{2}, \quad \exp\left(-\frac{(k-1)^2}{2n}\right) \ge a_k, \ k \ge 1. \tag{4.13}$$

As $\rho < 1$ so, using the formula (4.13), obtain

$$R_n(\rho) \le \sum_{1 \le k \le n/2} \rho^k \exp\left(-\frac{(k-1)^2}{2n}\right) \le \rho \int_0^{n/2} \rho^t \exp\left(-\frac{t^2}{2n}\right) dt,$$

$$R_n(\rho^{-1}) \ge \sum_{1 \le k \le n/2} \rho^{-k} \exp\left(-\frac{ck^2}{2n}\right) \ge$$

$$\ge \rho \int_1^{n/2} \rho^{-t} \exp\left(-\frac{ct^2}{2n}\right) dt - \exp\left(\frac{n\ln^2\rho}{2c}\right),$$

$$T_n(\rho) \le \sum_{n/2 < k \le n} \rho^k \exp\left(-\frac{(k-1)^2}{2n}\right) \le$$

$$\le \rho \int_{n/2-1}^{n} \rho^t \exp\left(-\frac{t^2}{2n}\right) dt, \quad T_n(\rho^{-1}) \ge 0.$$

Representing ρ in the form $\rho = \exp(-\alpha_n)$ and using the designations, obtain

$$R_n(\rho) = R_n^-, \quad R_n(1/\rho) = R_n^+, \quad T_n(\rho) = T_n^-, \quad T_n(1/\rho) = T_n^+.$$

Rewrite the last formulas in the form

$$R_n^- \le \exp(-\alpha_n) \int_0^{n/2} \exp\left(-\frac{t^2}{2n} - t\alpha_n\right) dt = R_{n,1}, \tag{4.14}$$

$$R_n^+ \ge \int_1^{n/2} \exp\left(-\frac{ct^2}{2n} + t\alpha_n\right) dt - \exp\left(\frac{\alpha_n^2 n}{2c}\right) = R_{n,2}, \tag{4.15}$$

$$T_n^- < e^{-\alpha_n} \int_{\frac{n}{2}-1}^{n} \exp\left(-\frac{t^2}{2n} - t\alpha_n\right) dt = T_{n,1}, \ T_n^+ < 0. \tag{4.16}$$

Denote $J_0(n) = \alpha_n\sqrt{n/2}$, $I_0(n) = \sqrt{n/8} + J_0(n)$ and accordingly to the formula

(4.14) obtain

$$R_n^- \leq R_{n,1} = e^{-\alpha_n} \int_0^{n/2} \exp\left(-\left(\frac{t}{\sqrt{2n}}\right)^2 - \right.$$

$$\left. -2\left(\frac{t}{\sqrt{2n}}\right)\frac{\sqrt{2n}\alpha_n}{2} - \frac{2n\alpha_n^2}{4} + \frac{2n\alpha_n^2}{4}\right) dt =$$

$$= e^{-\alpha_n} \int_0^{n/2} \exp\left(\frac{n\alpha_n^2}{2}\right) \exp\left(-\left(\frac{t}{\sqrt{2n}} + \sqrt{\frac{n}{2}}\,\alpha_n\right)^2\right) \frac{1}{\sqrt{2n}}\sqrt{2n}\, dt =$$

$$= e^{-\alpha_n} \int_{J_0(n)}^{I_0(n)} \exp\left(\frac{n\alpha_n^2}{2}\right) e^{-u^2}\, du\sqrt{2n} \leq$$

$$\leq \exp\left(-\alpha_n + \frac{n\alpha_n^2}{2}\right)\sqrt{2n}\int_{J_0(n)}^{\infty} e^{-u^2}\frac{2u \cdot du}{2u} \leq$$

$$\leq \sqrt{2n}\exp\left(-\alpha_n + \frac{n\alpha_n^2}{2}\right)\frac{\exp\left(-n\alpha_n^2/2\right)}{\sqrt{2n}\alpha_n} = \frac{e^{-\alpha_n}}{\alpha_n}. \qquad (4.17)$$

Denote

$$I_1(n) = \sqrt{cn/8} - \alpha_n\sqrt{n/2c}, \quad J_1(n) = \sqrt{c/2n} - \alpha_n\sqrt{n/2c}.$$

Then accordingly to (4.15) $R_n^+ \geq R_{n,2} = R'_{n,2} - R''_{n,2}$, where

$$R'_{n,2} = \int_1^{n/2} \exp\left(-\left(\sqrt{\frac{c}{2n}}t\right)^2 + 2\sqrt{\frac{c}{2n}}t\alpha_n\frac{1}{2}\sqrt{\frac{2n}{c}} - \frac{n\alpha_n^2}{2c} + \frac{n\alpha_n^2}{2c}\right) dt =$$

$$= \exp\left(\frac{n\alpha_n^2}{2c}\right)\int_1^{n/2} \exp\left(-\left(\sqrt{\frac{c}{2n}}t - \sqrt{\frac{2n}{c}}\right)^2\right) dt\frac{\sqrt{c/2n}}{\sqrt{c/2n}} =$$

$$= \exp\left(\frac{n\alpha_n^2}{2c}\right)\sqrt{\frac{2n}{c}}\int_{J_1(n)}^{I_1(n)} e^{-u^2}\, du = \sqrt{\frac{2n}{c}}\exp\left(\frac{n\alpha_n^2}{2c}\right)\int_{J_1(n)}^{I_1(n)} e^{-u^2}\, du,$$

$$R''_{n,2} = \exp\left(\frac{n\alpha_n^2}{2c}\right),$$

that is

$$R_n^+ \geq \exp\left(\frac{n\alpha_n^2}{2c}\right)\left(\sqrt{\frac{2n}{c}}\int_{J_1(n)}^{I_1(n)} \exp(-u^2)du - 1\right). \qquad (4.18)$$

Denote

$$I_2(n) = \frac{n}{\sqrt{2n}} + \alpha_n\sqrt{\frac{n}{2}}, \quad J_2(n) = \frac{n/2 - 1}{\sqrt{2n}} + \alpha_n\sqrt{\frac{n}{2}}.$$

Then accordingly to (4.16) we have

$$T_n^- \leq T_{n.1} = e^{-\alpha_n} \int_{(n/2)-1}^n \exp\left(-\frac{t^2}{2n} - t\alpha_n\right) dt =$$

$$= e^{-\alpha_n} \int_{\frac{n}{2}-1}^n \exp\left(-\left(\frac{t}{\sqrt{2n}}\right)^2 - 2\frac{t}{\sqrt{2n}}\frac{\sqrt{2n}}{2}\alpha_n - \right.$$

$$\left. - \left(\frac{\sqrt{2n}}{2}\alpha_n\right)^2 + \left(\frac{\sqrt{2n}}{2}\alpha_n\right)^2\right) \frac{dt}{\sqrt{2n}}\sqrt{2n} =$$

$$= \exp\left(-\alpha_n + \frac{n\alpha_n^2}{2}\right)\sqrt{2n}\int_{(n/2)-1}^n \exp\left(-\left(\frac{t}{\sqrt{2n}} + \sqrt{\frac{n}{2}}\alpha_n\right)^2\right)\frac{dt}{\sqrt{2n}} =$$

$$= \sqrt{2n}\exp\left(-\alpha_n + \frac{n\alpha_n^2}{2}\right)\int_{J_2(n)}^{I_2(n)} e^{-u^2}\,du \leq$$

$$\leq \sqrt{2n}\exp\left(-\alpha_n + \frac{n\alpha_n^2}{2}\right)\frac{\exp(-n(1/4 - 1/n + \alpha_n^2)/2)}{(\sqrt{n/8} - 1/\sqrt{2n} + \alpha_n\sqrt{n/2})} \leq$$

$$\leq \frac{4\exp(-\alpha_n - n(1/4 - 1/n)/2)}{(1 - 2/n)}. \tag{4.19}$$

As the result obtain

$$T_n^- \to 0, \ n \to \infty, \tag{4.20}$$

because $\alpha_n \to 0$ for $n \to \infty$.

The formulas (4.17), (4.18) and the condition $\alpha_n = n^{-1/2+\varepsilon}$, $0 < \varepsilon < 1/2$ lead to

$$\frac{R_n^-}{R_n^+} \leq \frac{e^{-\alpha_n}}{\alpha_n\exp\left(n\alpha_n^2/2c\right)}\left(\frac{\sqrt{2n}}{c}\int_{J_1(n)}^{I_1(n)} e^{-u^2}\,du - 1\right)^{-1} \sim$$

$$\sim \frac{1}{\alpha_n\exp\left(n\alpha_n^2/2c\right)}\left(\frac{\sqrt{2n}}{c}\int_{-\infty}^\infty e^{-u^2}\,du - 1\right) \to 0, \ n \to \infty, \tag{4.21}$$

$$R_n^+ \geq \sqrt{\frac{2n}{c}}\int_{J_1(n)}^{I_1(n)} e^{-u^2}\,du - 1 \to \infty, \ n \to \infty. \tag{4.22}$$

Then form the formulas (4.20)–(4.22) obtain that

$$\frac{R_n^- + T_n^-}{R_n^+ + T_n^!} \leq \frac{R_n^- + T_n^-}{R_n^+} \to 0, \ n \to \infty,$$

and consequently

$$P_n(\rho) \to 1, \ n \to \infty. \tag{4.23}$$

Put now $m = m_n = n^{(1+\varepsilon)/2}$ and denote

$$U_n^+ = \sum_{1 \le k \le m_n} \rho^k a_k, \quad V_n^+ = \sum_{m_n < k \le n} \rho^k a_k,$$

$$U_n^- = \sum_{1 \le k \le m_n} \rho^{-k} a_k, \quad V_n^- = \sum_{m_n < k \le n} \rho^{-k} a_k,$$

$$J_3(n) = \frac{m_n}{\sqrt{2n}} - \alpha_n \sqrt{\frac{n}{2}}, \quad D_n = \exp(\alpha_n + n\alpha_n^2).$$

Accordingly to the formula (4.12)

$$V_n^+ \le V_n^- = \sum_{m_n < k \le n} \rho^{-k} a_k \le \sum_{m_n < k \le n} \rho^{-k} \exp\left(-\frac{(k-1)^2}{2n}\right) =$$

$$= \sum_{m_n < k \le n} \exp\left(k\alpha_n - \frac{(k-1)^2}{2n}\right) \le \int_{m_n}^{\infty} \exp\left(t\alpha_n - \frac{t^2}{2n} + \alpha_n\right) dt + D_n,$$

$$V_n^- \le e^{\alpha_n} \int_{m_n}^{\infty} \exp\left(-\left(\frac{t}{\sqrt{2n}}\right)^2 + 2\left(\frac{t}{\sqrt{2n}}\right)\frac{\sqrt{2n}}{2}\alpha_n - \left(\frac{\sqrt{2n}}{2}\alpha_n\right)^2 + \right.$$

$$\left. + \left(\frac{\sqrt{2n}}{2}\alpha_n\right)^2\right) dt + D_n = \exp\left(\alpha_n + \frac{n\alpha_n^2}{2}\right) \sqrt{2n} \int_{J_3(n)}^{\infty} e^{-u^2} du + D_n \le$$

$$\le \frac{n}{m_n} \exp\left(\alpha_n + \frac{n\alpha_n^2}{2} - \left(\frac{m_n}{\sqrt{2n}} - \sqrt{\frac{n}{2}}\alpha_n\right)^2\right) + D_n =$$

$$= n^{(1-\varepsilon)/2} \exp\left(\alpha_n + \frac{n^{-\varepsilon} - (n^{\varepsilon/2} - n^{-\varepsilon/2})^2}{2}\right) + D_n \to 1, \quad n \to \infty,$$

that is

$$0 \le V_n^+ \le V_n^- \to 1, \quad n \to \infty. \tag{4.24}$$

As $\rho^{m_n} \sim \rho^{-m_n} \sim 1$, $n \to \infty$, so

$$U_n^+ \sim U_n^-, \quad n \to \infty. \tag{4.25}$$

Denote

$$J_4(n) = \alpha_n \sqrt{\frac{n}{2}}, \quad I_4(n) = \frac{m_n}{\sqrt{2n}} + J_4(n)$$

and estimate

$$U_n^+ = \sum_{1 \le k \le m_n} \rho^k a_k = \sum_{1 \le k \le m_n} e^{-k\alpha_n} a_k \ge \sum_{1 \le k \le m_n} e^{-k\alpha_n} \exp\left(-\frac{(k-1)^2}{2n}\right) \ge$$

$$\ge e^{-\alpha_n} \int_0^{m_n} \exp\left(-t\alpha_n - \frac{t^2}{2n}\right) dt = e^{-\alpha_n} \int_0^{m_n} \exp\left(-\left(\frac{t}{\sqrt{2n}}\right)^2 - \right.$$

$$\left. -2\left(\frac{t}{\sqrt{2n}}\right)\frac{\sqrt{2n}}{2}\alpha_n - \left(\sqrt{\frac{n}{2}}\alpha_n\right)^2\right) dt = e^{-\alpha_n} \int_{J_4(n)}^{I_4(n)} e^{-u^2} du \sqrt{2n} \sim$$

$$\sim \sqrt{2n} \int_0^{\infty} e^{-u^2} du \to \infty, \quad n \to \infty,$$

that is

$$U_n^+ \to \infty, \quad n \to \infty. \tag{4.26}$$

As

$$P_n = \frac{1 + U_n^- + V_n^-}{1 + U_n^+ + V_n^+ + U_n^- + V_n^-},$$

so accordingly to the formulas (4.24)–(4.26) obtain

$$P_n(\rho) = \frac{1 + 1/(U_n^- + V_n^-)}{1 + 1/(U_n^- + V_n^-) + (U_n^+ + V_n^+)/(U_n^- + V_n^-)} \sim$$

$$\sim [1 + (U_n^+ + V_n^+)/(U_n^- + V_n^-)]^{-1} \sim (1 + U_n^+/U_n^-)^{-1} \sim \frac{1}{2}, \quad n \to \infty,$$

consequently the formula $P_n(\rho) \to 1/2$, $n \to \infty$, is true.

Theorem 4.3. *If* $\rho = \exp(\alpha_n)$, $\alpha_n = n^{-1/2+\varepsilon}$, $1/2 > \varepsilon > 0$ *then*

$$\lim_{n \to \infty} P_n = 0.$$

If $\rho = \exp(\alpha_n)$, $\alpha_n = n^{-1/2-\varepsilon}$, $\varepsilon > 0$ *then*

$$\lim_{n \to \infty} P_n = \frac{1}{2}.$$

Proof. Using the replacement $\rho \to \rho^{-1}$ and the formula (4.6) it is easy to reduce the theorem 4.3 proof to the theorem 4.2 proof.

So we obtain complete classification of $P_n(\rho)$ limit behavior for $n \to \infty$. This classification recognizes the counteraction effects in the aggregated reserve system with the restoration and establishes the jump behavior of the limit $P_n(\rho)$, $n \to \infty$.

Consider now the results of the probability P_n numerical calculations for $\alpha_n = n^{-v}$ and for different v, n (fig. 7).

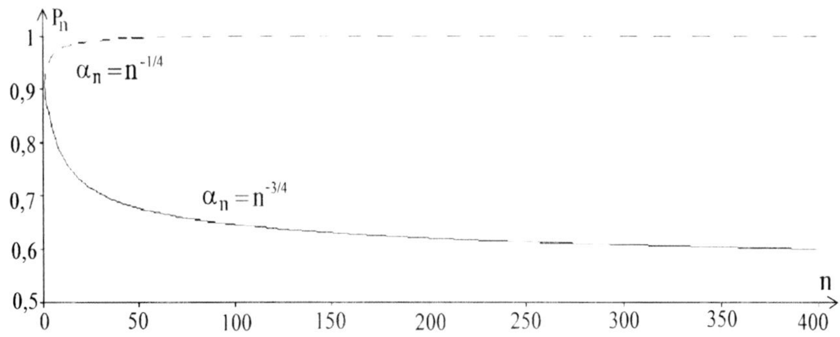

Fig. 7. The graphics of the dependence P_n on n for $v = 1/4,\ \ v = 3/4$.

$v \setminus n$	100	1000	2000	3000	4000
0,75	0,645	0,577	0,564	0,558	0,553
0,7	0,676	0,606	0,591	0,584	0,579
0,6	0,757	0,699	0,685	0,677	0,672
0,55	0,806	0,767	0,758	0,753	0,749
0,51	0,848	0,831	0,827	0,825	0,824
0,505	0,853	0,839	0,836	0,835	0,834
0,5	0,858	0,847	0,845	0,845	0,844
0,495	0,864	0,855	0,854	0,854	0,854
0,49	0,869	0,863	0,863	0,864	0,864
0,45	0,909	0,924	0,930	0,934	0,936
0,4	0,952	0,978	0.984	0,987	0,989
0,3	0,995	0,999	0,999	0,999	0,999
0,25	0,999	1	1	1	1

Table 1. The dependence of P_n on n for different v.

Obtained results (see the table 1) show that the convergence in the theorems 4.2, 4.3 to limit meanings 1, 1/2, 0 is slower if the parameter v is closer to the critical point 1/2.

4.2. Effectiveness of separate reserve

Consider now the quantitative estimates of the cooperative effects in some static reliability models. Consider n sequentially connected and independent elements with the reliability p, $0 < p < 1$. Then the reliability of this chain equals to p^n. Consider two alternative reserve schemes for the chain. The first scheme is the reserve of k parallel and independent chain duplicates. The reliability of this reserve scheme is

$$H^k = 1 - (1 - p^n)^k.$$

In the second scheme each element of the initial chain has its separate reserve with the volume k. The reliability of the second scheme is

$$H_k = (1 - q^k)^n, \quad q = 1 - p.$$

It is easy to show that $H_k \geq H^k$. But this inequality gives only qualitative representation about the separate reserve possibilities. To obtain the quantitative estimate it is convenient to replace the reliability by the necessary reserve volume.

Fix low bound $1 - \varepsilon$, $0 < \varepsilon < 1$ of the reliability and denote

$$k_1 = \min(k : H^k \geq 1 - \varepsilon), \quad k_2 = \min(k : H_k \geq 1 - \varepsilon).$$

For all a, $0 < a < 1$, the inequality

$$(1 - a)^m \geq 1 - ma, \quad m = 1, 2, \ldots \tag{4.27}$$

is true. Accordingly to the inequality

$$H^k \leq 1 - kp^n = kp^n,$$

obtain

$$k_1 \geq \frac{1 - \varepsilon}{p^n}. \tag{4.28}$$

Using the inequality (4.27) obtain

$$H_k \geq 1 - nq^k,$$

and so the formula

$$k_2 \leq \frac{|\ln(\varepsilon/n)|}{|\ln q|} + 1 \tag{4.29}$$

is true. The formulas (4.28), (4.29) comparison shows that for the large length n of the chain the scheme of the separate reserve has special advantages. It is worthy to mark that upper bound (which is logarithmical by n) of the necessary reserve volume may be easily spread from the initial chain onto the general case.

Really, suppose that G is connected and non-directed graph with the distinguished initial and final nodes. Suppose that this graph consists of n independent edges with the reliabilities $p_1, \ldots, p_n \geq 1 - q$, $0 < q < 1$. Denote by $H_G(p_1, \ldots, p_n)$ the reliability of this graph (the probability that there exists the way between initial and final nodes which is in the working state). As the graph G is connected so

$$H_G(p_1, \ldots, p_n) \geq p_1 \cdot \ldots \cdot p_n.$$

Define the graph G^k by the reservation of each edge with the volume k. Then the reliability $H_G^k(p_1, \ldots p_n)$ of the graph G^k satisfies the inequality

$$H_G^k(p_1, \ldots, p_n) \geq (1 - (1 - p_1)^k) \cdot \ldots \cdot (1 - (1 - p_n)^k) \geq (1 - q^k)^n \geq 1 - nq^k.$$

Consequently the quantity $k_G = \inf\{k : H_G^k(p_1, \ldots, p_n) \geq 1 - \varepsilon\}$ satisfies the formula

$$k_G \leq \frac{|\ln(\varepsilon/n)|}{|\ln q|} + 1. \tag{4.30}$$

So the logarithmical upper bound on n for the separate reserve scheme is true in general case.

Example of controlled electronic scheme

Consider the scheme with periodic impulse input signal

$$x(t) = \begin{cases} 1, & t \in A_i, \\ -1, & t \in B_i. \end{cases}$$

Here $A_i = C_{2i}$, $B_i = C_{2i+1}$, $i \geq 0$, with $C_k = [k, \ k + 1)$, $k \geq 0$. In a presence of stochastic disturbances this scheme may work during half interval C_k or do not work. Suppose that r.v. z_k equals 1 in the first case and equals 0 in the second case and r.v.'s z_k, $k \geq 0$, are independent with the distribution

$$P\{z_{2n+1} = 1\} = p_1, \ P\{z_{2n+1} = 0\} = 1 - p_1,$$

$$P\{z_{2n} = 1\} = p_2, \; P\{z_{2n} = 0\} = 1 - p_2, \; n \in \{0, 1, ...\}.$$

Denote by $q \in [0, 1]$ some control parameter of this scheme. A specifics of this system is that the function $p_1 = p_1(q)$ is monotonically increasing and the function $p_2 = p_2(q)$ is monotonically decreasing. This circumstance for long period was obstacle to reach necessary reliability of this scheme. And only the application of the cooperative effects allowed to get over this difficulty.

Design solution of this problem is in the replacement of the initial electronic scheme by two identical schemes. The input signal of the first scheme consists of the initial scheme negative component of the signal and the input signal of the second one consists of the positive component of the initial signal. Then it is possible to choose the control parameter q for each scheme separately. For example for the first scheme $q = 1$ and for the second one $q = 0$. Estimate the result of such design solution.

For this aim take the objective function which is usually used in technical reliability theory [1]:

$$p_N = P\{z_k = 1, \; k = 0, 1, \ldots, 2N - 1\}.$$

This function equals the probability of the reliable work during N steps. Accordingly to this choice of the parameter q we have

$$p_N = (p_1(1)p_2(0))^N.$$

Compare this quantity with the analogous quantity p'_N calculated for initial scheme (without separate choice of the parameter q),

$$p'_N = (\max_{q \in [0,1]} p_1(q)p_2(q))^N.$$

It is clear that

$$a_N = \frac{p_N}{p'_N} = \lambda^N, \tag{4.31}$$

where

$$\lambda = [p_1(1)p_2(0)]/[\max_{q \in [0,1]} p_1(q)p_2(q)] > 1.$$

The formula (4.31) gives the quantitative estimate of the cooperative effect in this example. It shows that the advantage of the separate choice of the parameter q increases as geometric progression by the time N. It is clear that the demand of the functions $p_1(q)$, $p_2(q)$ monotonicity is not necessary to obtain this effect. This demand may only to influence on the coefficient λ of the analyzed geometric progression.

In this problem the solution was found at first heuristically. Only after tests confirmation this technical solution was analyzed by means of the stochastic modelling [1].

Chapter 5

New approaches to data processing problems

The cooperative effects in the queueing, reliability and insurance models considered in the previous chapters create a source for new approaches and algorithms in data processing. A choice of the material of this chapter is defined by the following characteristics. At first, we consider the data processing problems which occurred in (complicated for traditional approaches) numerical experiments. Then, in the suggested solutions initial observations are transformed using decomposition methods. And lastly properties of obtained estimates of unknown parameters are investigated by the majorants and minorants of their quality indexes.

5.1. Growth models

In this subsection, statistical estimates of the parameters in the logistic and the Rikker growth models with random errors of observations are constructed. These models are interesting from a statistical viewpoint because they are based on non-linear recurrent formulas and so the method of the least squares (as the numerical experiment showed) gives sufficiently large relative error.

The first unknown (growth) parameter is expressed via the means of observed process trajectories "infinite" time intervals. Then these means are estimated via the observations and the parameters of the observation errors at the finite time intervals. If the time intervals tend to infinity then the obtained estimates tend to the accuracy meanings of the parameter by the probability. Results of numerical experiments showed that such algorithm of the parameter estimate is much more accuracy than the method of least squares.

Logistic growth model

Recurrent model of logistic growth

$$x_0 = a, \ x_{n+1} = bx_n(1 - x_n), \ n = 0, 1, ..., \tag{5.1}$$

where the parameters a, b satisfy the conditions

$$0 < a < 1, \ 1 < b < 4, \tag{5.2}$$

attracts large attention of physicists and biologists. It is connected with an appearance of pseudo stochastic regimes of the sequence x_n, $n = 0, 1, ...$, for some meanings of the (growth) parameter b. Recurrent model (5.1) may be is the simplest among the models creating pseudo stochastic regimes.

For this model it is interesting as from theoretical so from practical viewpoints to estimate the parameter b by inaccuracy observations y_n of the states x_n, $n = 0, 1, ...$ As the recurrent formula (5.1) is non linear so an application of least square method to the parameter estimate (widely used in linear systems [65]) is unnatural. This viewpoint is confirmed by manifold computer experiments made by N.A. Solopov. Intuitively it is clear that the quality properties of the sequence (5.1), like the existence of the limit cycles or the distributions dependently on the parameter b [70, p. 8–13], look more adequate. Our aim is to use these properties of the sequence x_n for the parameter b estimates by the inaccuracy observations y_n, $n = 0, 1, ...$

Denote that the condition (5.2) leads that the sequence (5.1) satisfies the inequalities

$$0 < x_n < 1, \ n = 0, 1, ... \tag{5.3}$$

Say that the sequence x_n, $n = 0, 1, ...$, has the limit cycle $x^{(1)}, ..., x^{(q)}$ with the length q, $q \geq 1$, if

$$\lim_{k \to \infty} x_{qk+j} = x^{(j)}, \ j = 1, ..., q. \tag{5.4}$$

Suppose that $p(dx)$ is the probability measure on the σ−algebra of Lebesgue measurable subsets of the set $[0, 1]$. Say that $p(dx)$ is the limit distribution of the sequence

x_n. $n = 0, 1, ...$, if for any Lebesgue measurable subset $C \subseteq [0, 1]$ we have

$$\lim_{n \to \infty} \frac{k(C, n)}{n} = \int_C p(dx) = p(C). \tag{5.5}$$

Here $k(C, n)$ is the number of x_i, satisfying the inclusion $x_i \in C$. $i = 0, 1, ..., n - 1$.

Denote:

$$X_n = \sum_{i=0}^{n-1} \frac{x_i}{n}, \quad X'_n = \sum_{i=0}^{n-1} \frac{x_i^2}{n}.$$

Say that the sequence x_n, $n = 0, 1, ...$, for fixed b satisfies the condition **(A)** if there exist the limits

$$\lim_{n \to \infty} X_n = \overline{x}, \quad \lim_{n \to \infty} X'_n = \overline{x^2}. \tag{5.6}$$

We shall consider additive

$$y_n = x_n + \epsilon_n \tag{5.7}$$

and product

$$y_n = x_n e^{\epsilon_n} \tag{5.8}$$

form errors of observations, $n = 0, 1, ...$. Here ϵ_n, $n = 0, 1, ...$ is the sequence of i.i.d.r.v.'s with common Gaussian d.f. (with the mean 0 and the variance σ^2.

Theorem 5.1. *If the sequence x_n, $n = 0, 1, ...$, has the limit cycle $x^{(1)}, ..., x^{(q)}$ with the length $q > 0$ or the limit distribution $p(dx)$ then it satisfies the condition* **(A)**. *If there is the limit cycle then*

$$\overline{x} = \sum_{j=1}^{q} \frac{x^{(j)}}{q}, \quad \overline{x^2} = \sum_{j=1}^{q} \frac{(x^{(j)})^2}{q}, \tag{5.9}$$

if there is the limit distribution then

$$\overline{x} = \int_0^1 x p(dx), \quad \overline{x^2} = \int_0^1 x^2 p(dx). \tag{5.10}$$

Proof. The formula (5.9) in the case of the limit cycle existence is obvious. Suppose that the sequence x_n, $n = 0, 1, ...$, has the limit distribution $p(dx)$ and define $\overline{x}$, $\overline{x^2}$ by the equalities (5.10). If for any natural number m there exists n_m so that for $n \geq n_m$ the inequalities

$$|X_n - \overline{x}| \leq \frac{2}{m}, \quad |X'_n - \overline{x^2}| \leq \frac{6}{m}, \tag{5.11}$$

are true then the sequence x_n, $n = 0, 1, \ldots$ has the property **(A)**.

To choose n_m divide the set $[0, 1]$ into the nonintersecting subsets $C_1 = [0, \frac{1}{m}]$, $C_2 = (\frac{1}{m}, \frac{2}{m}]$, $C_3 = (\frac{2}{m}, \frac{3}{m}]$, $\ldots$, $C_m = (\frac{m-1}{m}, 1]$. As $p(dx)$ is the limit distribution then from the formula (5.5) obtain that there exists natural number $n_m > 0$ so that for $n \geq n_m$ the inequalities

$$\left| \frac{k(C_i, n)}{n} - p(C_i) \right| \leq \gamma_m = \frac{1}{m^2}, \quad i = 1, \ldots, m \tag{5.12}$$

are true. From the formula (5.12) obtain that for $n \geq n_m$ the inequalities

$$\sum_{i=1}^{m} (p(C_i) - \gamma_m) \frac{(i-1)}{m} \leq X_n \leq \sum_{i=1}^{m} (p(C_i) + \gamma_m) \frac{i}{m}, \tag{5.13}$$

$$\sum_{i=1}^{m} (p(C_i) - \gamma_m) \frac{(i-1)^2}{m^2} \leq X'_n \leq \sum_{i=1}^{m} (p(C_i) + \gamma_m) \frac{i^2}{m^2} \tag{5.14}$$

are true. Replacing the summation by the integration in the formulas (5.13), (5.14) and using the formulas (5.10), obtain

$$\overline{x} - \frac{1}{m} - \frac{\gamma_m(m-1)}{2} \leq X_n \leq \overline{x} + \frac{1}{m} + \frac{\gamma_m(m+1)}{2}, \tag{5.15}$$

$$\overline{x^2} - \frac{2}{m} - \frac{\gamma_m m}{3} \leq X'_n \leq \overline{x^2} + \frac{(2 + \frac{1}{m})}{m} + \frac{\gamma_m m(1 + \frac{1}{m})^3}{3}. \tag{5.16}$$

The inequalities (5.15), (5.16) lead to

$$|X_n - \overline{x}| \leq \frac{1}{m} + \frac{\gamma_m(m+1)}{2} \leq \frac{1}{m} + \gamma_m m = \frac{2}{m},$$

$$|X'_n - \overline{x^2}| \leq \frac{(2 + \frac{1}{m})}{m} + \frac{\gamma_m m(1 + \frac{1}{m})^3}{3} \leq \frac{3}{m} + \frac{8\gamma_m m}{3} \leq \frac{6}{m} i.$$

So the inequalities (5.11) are proved.

Corollary 5.1. *If the theorem 5.1 conditions are true then*

$$0 < \overline{x^2} < \overline{x}, \tag{5.17}$$

$$b = \frac{\overline{x}}{\overline{x} - \overline{x^2}}. \tag{5.18}$$

Proof. From the conditions (5.2) obtain that the point 0 is not The limit cycle of the sequence x_n, $n = 0, 1, ...$, with the length 1 [70, p. 9]. So if the sequence x_n, $n \geq 0$, has limit cycle then for the equalities (5.9) obtain the formula (5.17).

Suppose now that for fixed b the sequence x_n, $n = 0, 1, ...$, has the limit distribution $p(dx)$. As this distribution is not concentrated at the point 0, that is $p(0) < 1$, then from the formulas (5.10) obtain the inequality (5.17). The equality $p(0)=1$ means that for some $n>0$ $k(0, n)>0$. Last statement is in the contradiction with the inequality (5.3). So the formula (5.17) is proved. The formula (5.1) leads to

$$\sum_{k=0}^{n-1} \frac{x_{k+1}}{n} = b \left(\sum_{k=0}^{n-1} \frac{(x_k - x_k^2)}{n} \right). \tag{5.19}$$

Transiting in the equality (5.19) to the limit for $n \to \infty$ and using the formulas (5.6) and (5.17), obtain (5.18).

Remark 1. *In the paper [70] it is shown that for all a, $1 < a < 4$, the sequence (5.1) has the limit cycle or the limit distribution and so the sequence satisfies the condition* **(A)**.

Begin to estimate the parameter b by the observations y_n. At first consider the additive model (5.7) of the observation error. Estimate the parameter b by the quantity

$$\hat{b}_n = \frac{Y_n}{Y_n - (Y_n' - \sigma^2)}. \tag{5.20}$$

Denote by ρ the mean square metric on the set of real random variables:

$$\rho(x, y) = (M(x - y)^2)^{1/2}.$$

Theorem 5.2. *If the theorem 5.1 conditions and the equalities (5.7) are true then there is the following convergence by the probability of $\hat{b}_n$ to b for $n \to \infty$.*

Proof. Calculate the means and the variances of r.v.'s Y_n, Y_n' in the limit $n \to \infty$:

$$MY_n = \frac{1}{n} \sum_{i=0}^{n-1} M(x_i + \epsilon_i) = X_n \to \overline{x},$$

$$DY_n = \frac{1}{n^2} \sum_{i=0}^{n-1} D(x_i + \epsilon_i) = \frac{1}{n^2} \sum_{i=0}^{n-1} D\epsilon_i = \frac{\sigma^2}{n} \to 0,$$

$$MY'_n = \frac{1}{n}\sum_{i=0}^{n-1} M(x_i + \epsilon_i)^2 = \frac{1}{n}\sum_{i=0}^{n-1}(x_i^2 + \sigma^2) \to \overline{x^2} + \sigma^2,$$

$$DY'_n = \frac{1}{n^2}\sum_{i=0}^{n-1} D(x_i + \epsilon_i)^2 = \frac{1}{n^2}\sum_{i=0}^{n-1} D(2x_i\epsilon_i + \epsilon_i^2) =$$

$$= \frac{1}{n^2}\sum_{i=0}^{n-1} M(2x_i\epsilon_i + \epsilon_i^2 - \sigma^2)^2 \le \frac{2}{n^2}\sum_{i=0}^{n-1}(D2x_i\epsilon_i + D\epsilon_i^2) =$$

$$= \frac{2}{n^2}\sum_{i=0}^{n-1}(4x_i^2 D\epsilon_i + D\epsilon_i^2) \le \frac{2}{n^2}\sum_{i=0}^{n-1}(4D\epsilon_i + D\epsilon_i^2) = \frac{4}{n}(4\sigma^2 + \sigma^4) \to 0.$$

From the last formulas and the triangle inequality obtain the convergence for $n \to \infty$:

$$\rho(Y_n, \overline{x}) \le \rho(Y_n, MY_n) + \rho(MY_n, \overline{x}) \to 0,$$

$$\rho(Y'_n, \overline{x^2} + \sigma^2) \le \rho(Y'_n, MY'_n) + \rho(MY'_n, \overline{x^2} + \sigma^2) \to 0. \tag{5.21}$$

Using the Chebyshev inequalities and the formulas (5.21) obtain the convergence by the probability

$$Y_n \to \overline{x}, \;\; Y'_n \to \overline{x^2} + \sigma^2, \;\; n \to \infty. \tag{5.22}$$

Combining the formulas (5.18), (5.20), (5.22), obtain the convergence by the probability of $\hat{b}_n$ to b for $n \to \infty$.

Consider now the product form (5.8) of the observations y_n, $n = 0, 1, ...$, errors. Choose the following estimate of the parameter b

$$\tilde{b}_n = \frac{Y_n e^{-\frac{\sigma^2}{2}}}{Y_n e^{-\frac{\sigma^2}{2}} - Y'_n e^{-2\sigma^2}}. \tag{5.23}$$

Theorem 5.3. *From the theorem 5.1 conditions and the equalities (5.8) obtain the convergence by the probability of $\tilde{b}_n$ to b for $n \to \infty$.*

Proof. Calculate the means and the variances of r.v.'s Y_n, Y'_n in the limit for $n \to \infty$:

$$MY_n = \frac{1}{n}\sum_{i=0}^{n-1} x_i M e^{\epsilon_i} = X_n e^{\frac{\sigma^2}{2}} \to \overline{x} e^{\frac{\sigma^2}{2}},$$

$$DY_n = \frac{1}{n^2}\sum_{i=0}^{n-1} x_i^2 D e^{\epsilon_i} = \frac{X'_n(e^{2\sigma^2} - e^{\sigma^2})}{n} \to 0,$$

$$MY_n' = \frac{1}{n}\sum_{i=0}^{n-1} x_i^2 M e^{2\epsilon_i} = X_n e^{2\sigma^2} \to \overline{x^2} e^{2\sigma^2},$$

$$DY_n' = \frac{1}{n^2}\sum_{i=0}^{n-1} x_i^4 D e^{2\epsilon_i} \le \frac{1}{n^2}\sum_{i=0}^{n-1} D e^{2\epsilon_i} = \frac{e^{8\sigma^2} - e^{4\sigma^2}}{n} \to 0.$$

From obtained formulas and the triangle inequality for $n \to \infty$ obtain the convergence

$$\rho(Y_n, \overline{x}e^{\frac{\sigma^2}{2}}) \le \rho(Y_n, MY_n) + \rho(MY_n, \overline{x}e^{\frac{\sigma^2}{2}}) \to 0,$$

$$\rho(Y_n', \overline{x^2}e^{2\sigma^2}) \le \rho(Y_n'MY_n') + \rho(MY_n'\overline{x^2}e^{2\sigma^2}) \to 0.$$

So the Chebyshev inequalities lead the convergence by the probability for $n \to \infty$:

$$Y_n \to \overline{x}e^{\frac{\sigma^2}{2}}, \ \ Y_n' \to \overline{x^2}e^{2\sigma^2}. \tag{5.24}$$

Combining the formulas (5.18), (5.23), (5.24), obtain the convergence by the probability of $\tilde{b}_n$ to b for $n \to \infty$.

Remark 2. *The additive model of the errors for the observations y_n, $n = 0, 1, ...,$ is more adequate for physical observations. The product model is more adequate for the models of population dynamics.*

Rikker model

In the mathematical ecology Rikker model

$$x_{n+1} = ax_n e^{-bx_n}, \ n = 0, 1, \ldots, \ \ a > 0, \ b > 0, \ x_0 > 0, \tag{5.25}$$

is of large interest [25, 69]. In these papers there are regions of the parameter (a, b) in which the sequence x_n, $n = 0, 1, \ldots,$ has the limit cycle or the limit distribution. A large attention is paid to the dependence of the limit cycle length on the parameter (a, b).

Construct the statistical estimate of the parameter (a, b) in the Rikker model (5.25) by the observations y_n of x_n, $n = 0, 1, \ldots,$ if the error of the observation y_n has the product form (5.8). This model leads to sufficiently complicated statistical problem in which the least square method (as in the previous model) is not adequate.

At first express the unknown parameter via the trajectories means

$$\lim_{n\to\infty} \frac{1}{n} \sum_{k=0}^{n-1} f(x_k)$$

for different functions f of x_n. These functions are chosen heuristically. Then separated means are estimated via the observations y_n, $n = 0, 1, \ldots$, and the errors distribution parameters (5.8). As a result construct the estimates which converge to unknown parameter by the probability for $n \to \infty$.

Suppose that $a > 1$ and the sequence x_n, $n \geq 0$ has the limit cycle with the length $q > 1$ or the limit non-degenerated distribution and denote

$$\overline{f(x)} = \lim_{n\to\infty} \frac{1}{n} \sum_{i=0}^{n-1} f(x_i).$$

In these suggestions the parameter (a, b) satisfies the equalities

$$b = 2\frac{\overline{x \ln x} - \overline{x}\,\overline{\ln x}}{\overline{x^2} - \overline{x}^2}, \quad \ln a = b\overline{x}. \tag{5.26}$$

Using the formula (5.26), construct the estimate $(\widehat{a}_n, \widehat{b}_n)$ of the parameter (a, b) by the observations y_i, $i = 0, 1, \ldots, n-1$:

$$\widehat{b}_n = 2e^{d/2}\frac{(Y \ln Y)_n - dY_n - Y_n(\ln Y)_n}{(Y^2)_n e^{-d} - (Y_n)^2}, \quad \widehat{a}_n = \exp(\widehat{b}_n Y_n e^{d/2}).$$

Here

$$Y_n = \frac{1}{n} \sum_{i=0}^{n-1} y_i, \quad (Y^2)_n = \frac{1}{n} \sum_{i=0}^{n-1} y_i^2,$$

$$(Y \ln Y)_n = \frac{1}{n} \sum_{i=0}^{n-1} y_i \ln y_i, \quad (\ln Y)_n = \frac{1}{n} \sum_{i=0}^{n-1} \ln y_i.$$

In [74] the convergence by the probability

$$(\widehat{a}_n, \widehat{b}_n) \to (a, b), \quad n \to \infty,$$

in the model (5.25) is proved.

5.2. Optimal algorithm of discrete images reconstruction

In this subsection the reconstruction algorithm of the function, which is constant in the checks of the square net (using integrals of this function along some straight lines), is suggested. This algorithm demands three arithmetical operations to calculate a single value of the reconstructed function.

The reconstruction of the discrete images (RDI) is one of intensively developed directions in the numerical tomography. Detailed review and manifold references on this topic are in the article [14] and in the monograph [32, chapter 1, § 5]. Here it is marked an importance of RDI algorithms acceleration.

In this subsection special measurement organization and RDI algorithms are suggested so that the problem becomes in a solution of linear equations system with the unit matrix. So a number of necessary calculations becomes to coincide with a number of variables. This algorithm is unimproved as by its velocity so by the calculations volume.

Suppose that unknown function f-image is defined on the square U with the side a. The square is divided into n^2 square cells and is represented at the figure 8 (here $n = 2$). In the (i,j)−th cell the function f is equal to $y'_{i,j}$. At the cell boundary without the restriction of the generality it is possible to define the function f by zero. Suppose that l is the segment of the line with ends belonging to the square U boundary. Define $F(l)$ as the integral of the function f along the segment l :

$$F(l) = \int_l f \, dl.$$

Denote $a_n = a/(n\sqrt{1 + 1/n^2})$ and put $y_{i,j} = y'_{i,j} \cdot a_n$. Suppose that $y_{i,0} = 0, \quad i = 1, \ldots, n$, and the equalities

$$y_{i,j} = y_{i,j-1} + x_{i,j}, \quad i, j = 1, \ldots, n \qquad (5.27)$$

are true. Our problem is to choose the segments so that integrals along them allow to reconstruct the matrix $X = \|x_{i,j}\|_{i,j=\overline{1,n}}$ by the minimal number of the arithmetical operations.

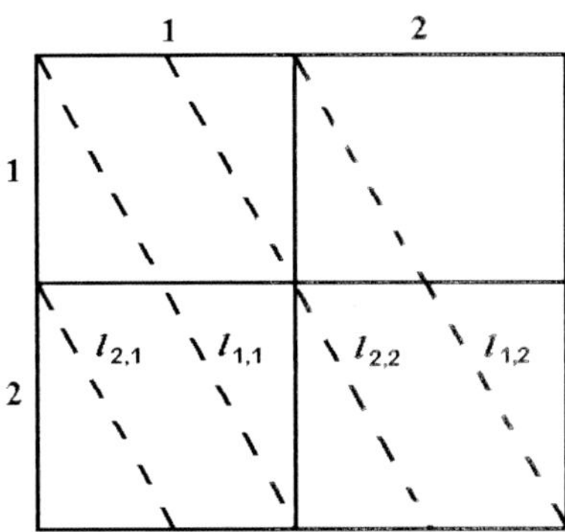

Fig. 8. The choice of the secants at the square network.

Define the segment $l_{1,1}$ by its final points: the left upper vertex of the cell $(1,1)$ and the right low vertex of the cell $(n,1)$. Then the segment $l_{i,j}$ passes through the left upper cell (i,j) parallel to the segment $l_{1,1}$. Denote

$$F_{i,j} = F(l_{i,j}), \quad i,j = 1,\ldots,n, \tag{5.28}$$

and put

$$F_{1,0} = 0, \ F_{n+1,j} = F_{1,j-1}, \quad j = 1,\ldots,n. \tag{5.29}$$

Then the equalities

$$x_{i,j} = F_{i,j} - F_{i,j+1}, \quad i,j = 1,\ldots,n, \tag{5.30}$$

are true.

So if the formulas (5.28), (5.29) are true then it is possible to define each element of the matrix X by the formula (5.30), using single arithmetical operation (of the subtraction). If there are n^2 parallel computing devices then this algorithm becomes the most fast. Total volume of the calculations n^2 which are necessary to define complete matrix X is minimal too. The matrix $\|y_{i,j}\|_{i,j=\overline{1,n}}$ may be calculated by the formulas (5.27), (5.30), using $2n^2$ arithmetical operations (by 2 operations per one element).

5.3. Layering method in biorhythmics problems

There is a problem of small data volumes in the data processing. One of the most effective method of its solution is so called layering method, widely used in quality control [67, p. 18–20]. This method occurred fruitful in manifold concrete problems. In this subsection some variant of this method is applied to the biorhythmics problems. The analytical foundation of this variant is suggested.

In the monograph [28, p. 47] 7479 histories and biorhythmics reports (with the established beginning hour by clinical indexes) of different illnesses (gastric ulcer, pancreatitis and etc.) are analyzed. The frequency curve of illnesses beginning moments with five pikes $A_1, ..., A_5$ is constructed. Our problem is to verify an accuracy of the hypothesis of five pikes presence.

The Chebyshev inequality gives the following formula

$$P(\max_{1 \leq i \leq m} |\widehat{p}_i - p_i| < 1/\sqrt{n\gamma}) \geq 1 - \gamma, \ \ 0 < \gamma < 1 \tag{5.31}$$

with n – the number of the observations (in our case $n = 7479$). Construct the confidence intervals for the theoretical probabilities p_i in the points of the local extremum of the frequency curve. The calculations by the formula (5.31) showed that the hypothesis (about the excess of the theoretical probabilities in the points of the frequency curve local maximum over the theoretical probabilities in the neighbor points of the local minimum is true with the probability 0.52. So this hypothesis (about five pikes) has the small consistence.

In this subsection the following ideas of the layering method verify the hypothesis about different pikes separately. From the initial frequency curve $\widehat{p}_i$, $1 \leq i \leq 24$, we extract the observations connected only with the local maximum. Using these observations consider five frequency curves connected with the separated pikes. To verify the hypothesis for each pike we use the analogs of the formula (5.31) for $m = 3$. The confidence intervals which confirm the hypothesis for the separate pikes are obtained with the probabilities equal to 0.9. The cause of the increasing of these probabilities (from 0.52 to 0.9) in the situation, when observations volumes decrease, may be explained as follows.

Consider the curve of the empirical probabilities (frequencies) $\widehat{p}_i = n_i/n, 1 \leq i \leq m$, and the corresponding curve of the theoretical probabilities $p_i, 1 \leq i \leq m$. Here n_i is the number of the observations with the meaning $i, 1 \leq i \leq m, n = \sum_{i=1}^{m} n_i$. Extract the observations connected with the points of the frequency curve between two neighbor local minimum $i_1, i_2, ..., i_{m'-1}, i_{m'}$ ($i_1 \geq 1, \ i_{m'} \leq m$). Construct then the curve of the empirical probabilities

$$\widehat{p'_{i_j}} = \frac{n_{i_j}}{n'}, \quad n' = \sum_{j=1}^{m'} n_{i_j}, \quad 1 \le j \le m'.$$

Correspond to this curve the appropriate curve of the theoretical probabilities

$$\pi'_j = \frac{p_{i_j}}{\pi}, \quad \pi = \sum_{j=1}^{m'} p_{i_j}, \quad 1 \le j \le m'.$$

Fix some $r, s,\ 1 \le r, s \le m'$ so that $\Delta_{r,s} = n_{i_r} - n_{i_s} > 0$ and consider the hypothesis $H_{r,s} = (p_{i_r} > p_{i_s})$. To confirm the hypothesis $H_{r,s}$ with the probability $(1 - \gamma)$ by the frequency curve $\widehat{p_i},\ 1 \le i \le m$, accordingly to the inequality (5.31) it is necessary that the inequality

$$\Delta_{r,s} \ge 2\sqrt{n/\gamma} \tag{5.32}$$

is true. To verify the hypothesis $H_{r,s}$ by the frequency curve $\widehat{p'_i},\ 1 \le i \le m'$, it is enough to weaken the inequality (5.32) to

$$\Delta_{r,s} \ge 2\sqrt{n'/\gamma}. \tag{5.33}$$

In our case $n' \sim 1000$. This fact in the connection with the inequalities (5.32), (5.33) explain the obtained effect.

5.4. Stationary random processes with finite correlation interval

To estimate parameters of random processes observed in the oceanology, the ocean acoustics and biorhythmics following statistical model [47,48] is used. The observed process $Y(t),\ MY(t) = 0$ is considered as the real Gaussian stationary process with the finite correlation interval A :

$$A = \sup(T : b(T) = MY(0)Y(T) \ne 0) > 0. \tag{5.34}$$

The process $Y(t)$ is divided into the stochastic Fourier integral [65, p. 215, 216] with the spectral density $f(\lambda),\ -\infty < \lambda < \infty$. The problem of the function $f(\lambda)$ reconstruction by the process $Y(t)$ is considered. Usually it is solved by the known formula

$$f(\lambda) = \frac{1}{2\pi} \int_{-\infty}^{\infty} e^{-i\lambda t} b(t)\,dt.$$

The spectral density $f(\lambda)$ estimate in the single point demands an approximate knowledge of the function $b(t)$ but in sufficiently large number of real axis points. Another disadvantage of the process $Y(t)$ representation in the form of the stochastic Fourier integral for the model (5.34) is a large complexity of a stochastic simulation even in the simplest versions of this model. So there is a problem of a convenient (for the computer simulation) representation of the process $Y(t)$.

Consider the process $Y(t)$, representing by the rows

$$Y(t) = \sum_{i=0}^{\infty} a_i y_i(s^i t), \quad -\infty < t < \infty. \tag{5.35}$$

Here $s, a_0, a_1, \ldots$ satisfy the conditions

$$s > 1, \quad A = \sum_{i=0}^{\infty} A_i < \infty, \quad A_0 = a_0^2, \quad A_1^2 = a_1^2, \ldots \tag{5.36}$$

and $y_0(t)$, $y_1(t), \ldots$ are independent and identically distributed Gaussian processes, which satisfy the condition

$$B(T) = M y_0(0) y_0(T) = \max(0, 1 - |T|), \quad -\infty < T < \infty. \tag{5.37}$$

The processes (5.35), (5.36), (5.37) satisfy the condition (5.34). The simplest computer simulation of the processes $y_0(t)$, $y_1(t), \ldots$ may be defined by the function

$$F(T) = M(Y(0) - Y(T))^2.$$

Reconstruct the numbers A_i by $D_j = F(s^{-j})$, $j = 0, 1, \ldots$, as follows

$$A_i = \frac{s(D_i - D_{i+1}) - (D_{i-1} - D_i)}{2(s-1)}, \quad i = 0, 1, \ldots, \tag{5.38}$$

where $D_{-1} = D_0$. Consider multi-dimension generalization of the formulas (5.35), (5.36), (5.37), (5.38).

One-dimension case

Suppose that $X = \{x_k, \ k \in Z\}$ is the sequence of independent r.v.'s which have the Gaussian distribution $\Phi_{0,1}(t)$ with the mean 0 and the unit variance. Suppose that r.v. τ is uniformly distributed on the set $[0, 1]$ r.v. and the random objects X, τ are independent. Put the random function $x(t)$ by the equalities

$$x(t) = x_k, \quad k \le t < k+1, \quad k \in Z \tag{5.39}$$

and suppose that

$$y_0(t) = x(t + \tau), \quad -\infty < t < \infty. \tag{5.40}$$

Using the equalities (5.39), (5.40), it is simple to prove that the random process $y_0(t)$ is Gaussian and stationary (that is its finite-dimension distributions of $y_0(t)$ are invariant to the argument t shift) and satisfy the equality (5.37). It is known that the spectral density $f_0(\lambda)$ of the random process $y_0(t)$ satisfies the equality

$$f_0(\lambda) = \frac{1 - \cos \lambda}{\pi \lambda^2}, \quad -\infty < \lambda < \infty.$$

Using the obvious equalities

$$My_i(t) \equiv 0, \quad My_i^2(t) \equiv 1,$$

calculate the function $F(T), T > 0$:

$$F(T) = M \left[\sum_{i=0}^{\infty} a_i (y_i(0) - y_i(s^i T)) \right]^2 =$$

$$= \sum_{i=0}^{\infty} A_i M (y_i(0) - y_i(s^i T))^2 = 2 \left(A - \sum_{i=0}^{\infty} A_i B(s^i T) \right).$$

From the formula (5.37) and the conditions (5.36) obtain

$$D_j = F(s^{-j}) = 2 \left(A - \sum_{i=0}^{j} A_i s^{i-j} \right),$$

$$\frac{D_j - D_{j+1}}{2(1 - 1/s)} = \sum_{i=0}^{j} A_i s^{i-j}; \quad j = 0, 1, \dots \tag{5.41}$$

The formula (5.38) is the corollary of (5.41).

Multidimensional case

For a simplicity we may consider the two-dimensional variant. Suppose that $\mathcal{X} = \{x_{k_1, k_2}, \ k_j \in Z, \ j = 1, 2\}$ is the sequence of independent r.v.'s with the Gaussian distribution $\Phi_{0,1}(t)$. And (τ_1, τ_2) are uniformly distributed on the unique square $[0, 1] \times [0, 1]$ random vectors. The random objects $\mathcal{X}, \ (\tau_1, \tau_2)$ are independent. Define the random functions

$$x(t_1, t_2) = x_{k_1, k_2}, \quad k_j \le t_j < k_{j+1}, \quad k_j \in Z, \quad j = 1, 2,$$

$$y(t_1, t_2) = x(t_1 + \tau_1, t_2 + \tau_2), \quad -\infty < t_1, t_2 < \infty.$$

It is easy to prove that the random function $y(t_1, t_2)$ is stationary and Gaussian and the equalities

$$My(0,0)y(T_1, T_2) = \prod_{j=1}^{2} B(T_j) \tag{5.42}$$

are true.

Suppose that $y_{i_1,i_2}(t_1, t_2)$, $i_1, i_2 = 0, 1, \ldots$, are the independent random "copies" of the random function $y(t_1, t_2)$ and positive numbers s_1, s_2, a_{i_1,i_2}, $i_1, i_2 = 0, 1, \ldots$, satisfy the conditions

$$s_1 > 1, \quad s_2 > 1, \quad a_{i_1,i_2}^2 = A_{i_1,i_2}, \quad \sum_{i_1=0}^{\infty} \sum_{i_2=0}^{\infty} A_{i_1,i_2} = A < \infty. \tag{5.43}$$

Then the gaussian stationary function

$$Y_{t_1,t_2} = \sum_{i_1=0}^{\infty} \sum_{i_2=0}^{\infty} a_{i_1,i_2} y_{i_1,i_2}(s_1^{i_1} t_1, s_2^{i_2} t_2)$$

satisfies the equality

$$F(T_1, T_2) = M(Y(0,0) - Y(T_1, T_2))^2 =$$

$$= 2(A - \sum_{i_1-0}^{\infty} B(s_1^{i_1} T_1) \sum_{i_2-0}^{\infty} A_{i_1,i_2} B(s_2^{i_2} T_2)) \tag{5.44}$$

and so has the unit square of the correlation.

Denote $D_{i_1,i_2} = F(s_1^{-i_1}, s_2^{-i_2})$ and consider the problem of $\{A_{i_1,i_2}\}$ reconstruction by the set $\{D_{i_1,i_2}\}$. From the formula (5.44) obtain

$$-\frac{1}{2} D_{j_1,j_2} + A = \sum_{i_1=0}^{j_1} s_1^{i_1-j_1} \mathcal{A}_{i_1,j_2} \tag{5.45}$$

where

$$\mathcal{A}_{i_1,j_2} = \sum_{i_2=0}^{j_2} A_{i_1,i_2} s_2^{i_2-j_2}, \quad j_1, j_2 = 0, 1, \ldots.$$

Using the formula (5.38), find the solution of the system (5.45):

$$\frac{s_1(D_{i_1,j_2} - D_{i_1+1,j_2}) - (D_{i_1-1,j_2} - D_{i_1,j_2})}{2(s_1 - 1)} = -\frac{1}{2} C_{i_1,j_2} = \mathcal{A}_{i_1,j_2} \tag{5.46}$$

where $i_1. j_2 = 0, 1, \ldots,$ $D_{-1,j_2} = D_{0,j_2}$. Putting $C_{i_1,-1} = C_{i_1,0}$ and again using the formulas (5.38), (5.46), find

$$\frac{s_1\left(C_{i_1,i_2} - C_{i_1,i_2+1}\right) - \left(C_{i_1,i_2-1} - C_{i_1,i_2}\right)}{2(s_2 - 1)} = A_{i_1,i_2}, \qquad (5.47)$$

$$i_1, i_2 = 0, 1, \ldots.$$

The replacement of the row (5.35) by the stochastic integral and the analogs of the formulas (5.38), (5.46), (5.47) for this case are given in [87].

5.5. Estimates of parameters in parallel and sequential connections of independent elements

In the model of the parallel connected independent elements (see fig. 9) the failure probability $\overline{P}$ is calculated via the failure probabilities $\overline{p}_1, \ldots, \overline{p}_n$ of the separate elements by the formula

$$\overline{P} = \overline{p}_1 \cdot \ldots \cdot \overline{p}_n, \quad \overline{p}_1 = 1 - p_1, \ldots, \overline{p}_n = 1 - p_n.$$

This model is interesting for a consideration of high reliable systems.

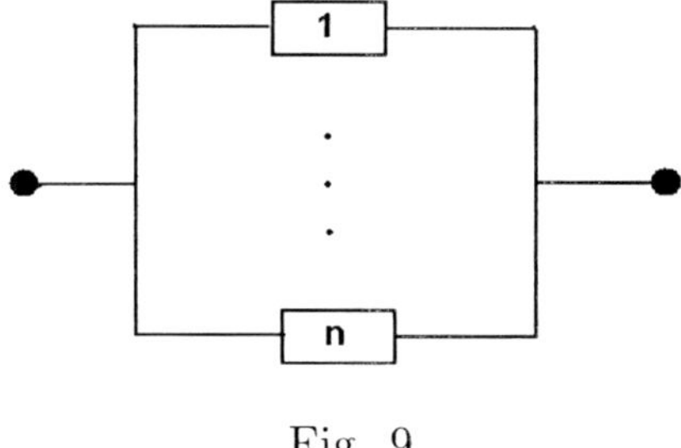

Fig. 9

Controversy in the model of sequential independent elements (see fig. 10)

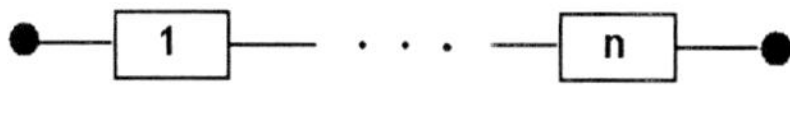

Fig. 10

the reliability P is calculated via the reliabilities $p_1, \ldots, p_n$ of the separate elements by the formula

$$P = p_1 \cdot \ldots \cdot p_n.$$

Consider only the scheme of the sequential independent elements because the scheme with the parallel elements may be considered analogously.

Compare two variants of the sequential scheme test. First of them is based on N independent tests of the complete scheme. Using the tests, it is possible to define the frequency $\widehat{P}$ of the scheme work in N tests. It is non deviated estimate of the parameter P. The second variant is in N independent tests of the separate elements. The tests define the frequencies $\widehat{p}_1, \ldots, \widehat{p}_n$ of the separate elements work. Then the non deviated estimate of the parameter P is constructed:

$$\widetilde{P} = \widehat{p}_1 \cdot \ldots \cdot \widehat{p}_n.$$

Compare the variances of these estimates if for some a, b, $0 < b \le a < 1$, the conditions

$$b \le p_1, \ldots, p_n \le a$$

are true. These conditions lead to the fast (with geometric progression rate) decreasing of the parameter P with the parameter N growth.

Suppose that the number N is sufficiently large comparably with n, that is

$$\frac{b^{-1} - 1}{N} \le \frac{\ln 2}{n},$$

then it is easy to obtain the formula

$$K_n - \frac{D\widehat{P}}{D\widetilde{P}} \ge \frac{(a^{-n} - 1)\ln 2}{n(b^{-1} - 1)}.$$

So there is the geometric by n rate of variances ratio K_n decreasing.

This result increases the statement of the paper [58] concerning the case of the fixed tests volume. It helps to understand better the global quality control system which is widely used in Japan [67].

5.6. Estimate of exponential distribution parameter

Consider now an estimate of the parameter T of exponential d.f.

$$F(t) = (1 - e^{-t/T}), \quad t \ge 0,$$

in the case of large T, $T \gg 1$. Such a distribution occurs in the queuing theory, the reliability theory and etc. It characterizes the moment of rare event appearance

[95, 73], for example a failure in high reliable system. So this problem is important for tests of new technologies and etc. And there is a necessity to construct fast schemes of tests and accuracy estimates of the parameter.

Suppose that $x_1, \ldots, x_N$ are independent failure times with d.f. $F(t)$. Then $\widehat{T} = (x_1 + \ldots + x_N)/N$ is non-deviated estimate of the parameter T with the variance

$$D\widehat{T}_N = \frac{T^2}{N}.$$

The mean time V_N of N primary tests satisfies the inequality $V_N \geq T$. To diminish V_N suppose that the number of primary tests is increased into m times and all these tests begin at the same moment 0. Denote $0 \leq t_1 \leq t_2 \leq \ldots \leq t_N$ moments of first N among Nm possible failures and suppose that

$$t_1 = y_1. \; t_2 - t_1 = y_2, \ldots, \; t_N - t_{N-1} = y_N.$$

Construct non-deviated estimate of the parameter T

$$\widetilde{T} = \frac{1}{N} \left(Nmy_1 + (Nm - 1)y_2 + \ldots + (Nm - N + 1)y_N \right)$$

with the variance T^2/N. Then the mean time U_N of this procedure, which coincides with the mean of t_N, satisfies the inequality $U_N \leq T/(m-1)$.

So, increasing in m times the tests general number, we conserve the mean and the variance of the parameter T estimate but decrease into $(m-1)$ times the total mean time of the tests.

5.7.   Recognition of extremal objects by interval decision rule

Suppose that $m-$dimension vectors $X_1 = (x_{11}, ..., x_{1m}), \ldots, \; X_n = (x_{n1}, ..., x_{nm})$ characterize the objects $1, ..., n$ by the factors $1, ..., m$. Divide the objects $1, \ldots, n$ into two sets $I_0. \; I_1$. First of them I_0 consists of so called extremal objects, and second set I_1 consists of non-extremal objects. Introduce the interval decision rule, which will recognize the extremal objects by the factors $1, \ldots, m$.

The object $X = (x_1, \ldots, x_m)$ is considered as extremal by the interval decision rule if

$$\min(x_{1i}, \ldots, x_{ni}) = m_i \leq x_i \leq M_i = \max(x_{1i}, \ldots, x_{ni}), \tag{5.48}$$

$$1 \leq i \leq m.$$

All extremal objects $i \in I_0$ are recognized as extremal ones by the decision rule (5.48). Denote by $n_{10} = n_{10}(m)$ the number of non-extremal objects $i \in I_1$, which satisfy the condition (5.48). These objects are recognized incorrectly. If the number m of the factors is increased to $m + 1$ then the number $n_{10}(m + 1) \leq n_{10}(m)$.

To calculate m_i, M_i, $1 \leq i \leq m$, and to calculate n_{10} it takes not more than $2nm$ arithmetical operations. This property allows to construct very economical recognition images procedure in a comparison for an example with the method of linear discriminant function and the method of dead-end tests.

A construction of the set I_0 in applications is based on the following procedure. Suppose that besides of the factors $1, \ldots, n$ we have so called main factor. That is there is additional factor 0 with the meanings $x_{10}, \ldots, x_{n0}$. Choose the critical level a for the factor 0 and define the set I_0 of the extremal objects by the condition

$$I_0 = \{i : \ 1 \leq i \leq n, \ x_{i0} \geq (\text{or} \ \leq) \ a\}. \tag{5.49}$$

Numerical experiments in the problems of extremal objects recognition (the extremal icing of Okhotskoe sea, the large breaks of the encephalitis in Primorye, the high catches of red fish in different regions of Russian Far East and etc.) showed that an increasing of the critical level leads usually to the significant decreasing of the number $n_{10}(m)$ (of the incorrectly recognized objects).

If an user needs to change the condition (5.49) or the number of main factors it may be easily realized.

Chapter 6

Cooperative effects in stochastic models of mechanics

In this chapter the cooperative effects in the following stochastic models are considered: anomalous diffusion at the interval with reflecting edges and periodical initial conditions, the break of threads wisp, the sliding regime. A specifics of these models is a constructive introduction to large parameters. Without this parameter, a quantitative investigation of these models becomes very complicated.

6.1. Anomalous Diffusion with Periodical Initial Conditions on Interval with Reflecting Edges

In [92] the mathematical model of the anomalous diffusion in a space was suggested. This model originates in an investigation of processes in complex systems with variable structure: glasses, liquid crystals, biopolymers, proteins and etc and a turbulence in a plasma [72].

In this model, a coordinate of a particle has stable distribution (not normal one). As a result, the density of its distribution function satisfies an analog of the diffusion equation in which second derivative by coordinate is replaced by a partial derivative.

In a comparison with [92] in this subsection the anomalous diffusion with periodic initial conditions on the interval with reflecting edges is considered. As A.A. Borovkov suggested for its investigation special probability methods are developed to analyze such the diffusion. In the case of normal diffusion, the constructed model has an interest for technical mechanics in an analysis of a fuel mixing in a straight flow engine [7].

Construction of Anomalous Diffusion Model
on the Interval $[-1,\ 1]$

Suppose that $y(t)$, $t \geq 0$, is the homogenous random process with independent increments and the initial condition $y(0) = 0$. The difference $y(t) - y(\tau)$, $t > \tau \geq 0$, has symmetric stable distribution on the straight line $(-\infty,\ \infty)$ with the parameter a, $0 < a \leq 2$, and the characteristic function

$$M \exp(iu[y(t) - y(\tau)]) = \exp(-(t - \tau)|u|^a).$$

The process $y(t)$, $t \geq 0$, describes [92] the anomalous diffusion on a straight line.

Each realization of the random process $y(t)$, $t \geq 0$ may be considered as the curve Γ on the plane (y, t) and may be represented in the parametrical form: $y = y(\tau)$, $t = t(\tau) = \tau$, $\tau \geq 0$.

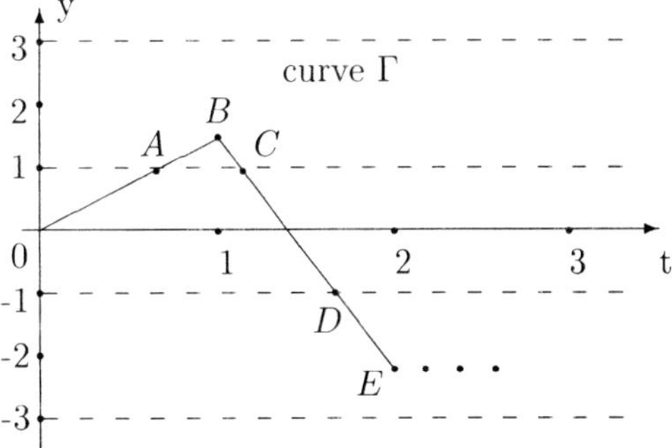

Fig. 11

Suppose that the curve Γ is reflected from the lines $y = 1$, $y = -1$. For this aim represent the plane $(y,\ t)$ as a transparent and infinitely thin sheet of a paper with the curve Γ. Bend this sheet of the paper along the lines $y = \pm 1$, $y = \pm 3, \ldots$ into the transparent strip $-1 \leq y \leq 1$ with fragments of initial curve Γ. As a result Γ is transformed into the curve γ (see fig. 12) : $y = Y(\tau)$, $t = t(\tau) = \tau$. Analogously to [92] the random process $Y(t)$, $t \geq 0$, may be considered as the model of the anomalous diffusion on the interval $[-1,\ 1]$ with reflecting edges. It is clear that if the curve Γ coincides with some straight line then the curve γ is constructed accordingly to the law of the geometrical optics: a falling angle equals to a reflecting angle.

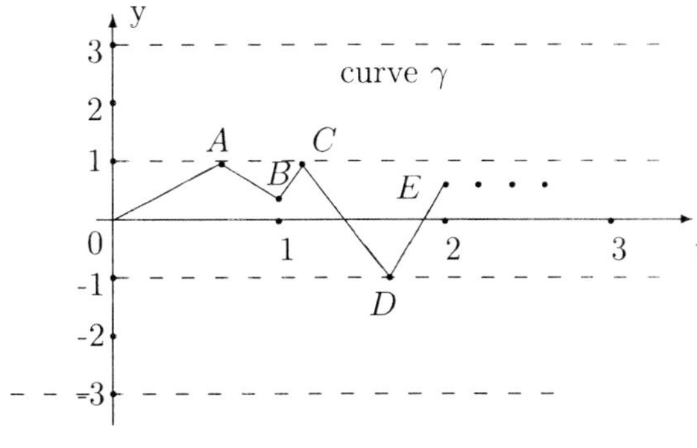

Fig. 12

Additionally to the geometrical representation of the process $Y(t)$, $t \geq 0$, consider its analytical representation by the functions $f : R \to [-2, 2)$, $g : [-2, 2) \to [-1, 1]$,

$$f(u) = (u + 2)/mod\ 4 - 2,$$

$$g(u) = \begin{cases} u, & -1 \leq u \leq 1, \\ 2 - u, & 1 < u < 2, \\ -2 - u, & -2 \leq u < -1, \end{cases}$$

$$Y(t) = g(f(y(t))). \tag{6.1}$$

Here $u/mod\ A = A\{u/A\}$, $A > 0$, and $\{z\}$ is the fractional part of the real number z. It is known from the group theory that for $u,\ v \in R,\ A > 0$

$$(u/mod\ A)/mod\ A = u/mod\ A,$$

$$(u + v)/mod\ A = (u/mod\ A + v/mod\ A)/mod\ A. \tag{6.2}$$

Define on the half-interval $[-2,\ 2)$ the binary operation "$\oplus$" and the unary operation of the inverse "$\ominus$" :

$$u \oplus v = f(u + v),\quad \ominus u = f(-u),\quad u,\ v \in [-2,\ 2).$$

Prove that these operations on the half-interval $[-2,\ 2)$ form the commutative group C with the summation $\oplus$ and the inverse $\ominus$ and the unit 0 :

$$u \oplus v = v \oplus u,\quad u \oplus (\ominus u) = 0,\quad (u \oplus v) \oplus w = u \oplus (v \oplus w),\quad u, v, w \in C.$$

Really using (6.2) obtain

$$u \oplus v = f(u+v) = (u+v+2)/mod\ 4 - 2 = (v+u+2)/mod\ 4 - 2 =$$

$$= f(v+u) = v \oplus u,$$

$$u \oplus (\ominus u) = (u + (\ominus u) + 2)/mod\ 4 - 2 = (u + (-u+2)/mod\ 4 - 2+$$

$$+2)/mod\ 4 - 2 = (u - u + 2)/mod\ 4 - 2 = f(0) = 0,$$

$$(u \oplus v) \oplus w = [f(u+v) + w + 2]/mod\ 4 - 2 = [(u+v+2)/mod\ 4 - 2 + w+$$

$$+2]/mod\ 4 - 2 = (u+v+w+2)/mod\ 4 - 2 = [u + (v+w+2)/mod\ 4 - 2+$$

$$+2]/mod\ 4 - 2 = [u + f(v+w) + 2]/mod\ 4 - 2 = u \oplus (v \oplus w).$$

Prove that $f : (R, +) \to (C, \oplus)$ is the homomorphism of the additive group R of the real numbers onto the group C :

$$f(u+v) = f(u) \oplus f(v), \quad f(-u) = \ominus f(u).$$

Using (6.2), obtain

$$f(u) \oplus f(v) = f(f(u) + f(v)) = [(u+2)/mod\ 4 - 2 + (v+2)/mod\ 4 - 2+$$

$$+2]/mod\ 4 - 2 = (u+v+2)/mod\ 4 - 2 = f(u+v),$$

the equality $f(-u) = \ominus f(u)$ is the corollary of the inverse $\ominus$ definition.

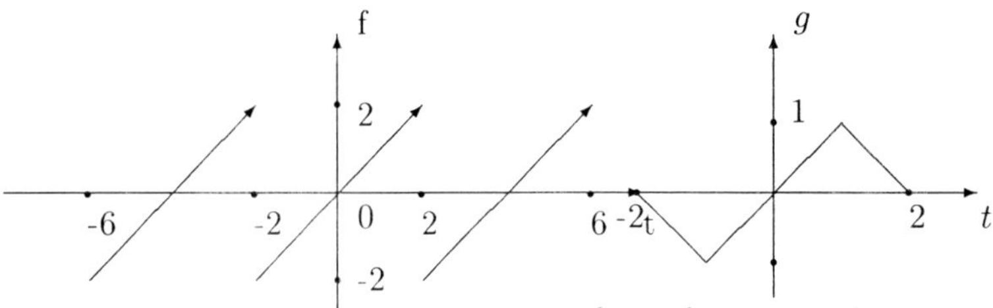

Fig. 13. The graphics of the functions f, g.

Distribution of reflected process and its convergence to uniform distribution

Denote by $p_t = p_t(u)$, $\pi_t = \pi_t(u)$, $P_t = P_t(u)$ the distribution densities of r.v.'s $y(t)$, $Y(t)$, $f(y(t))$, accordingly. Using the formula (6.1) and the graphic of the function f (see fig. 13) find:

$$P_t(u) = \sum_{v:\, f(v)=u} p_t(v) = \sum_{k=-\infty}^{\infty} p_t(u - 4k), \quad u \in [-2,\, 2), \qquad (6.3)$$

$P_t(u) = 0$, $u \notin [-2,\, 2)$. It is easy to prove from [24, the chapt. 17, §6, the lemma 1] that the row (6.3) converges.

Define the auxiliary function

$$\overline{P}_t(u) = \sum_{k=-\infty}^{\infty} p_t(u - 4k), \quad -\infty < u < \infty.$$

The function $\overline{P}_t(u)$ coincides with $P_t(u)$ for $u \in [-2,\, 2)$ and has the period 4 and is symmetric. So from the formula (6.1) and the graphic of the function g (see fig. 13) obtain

$$\pi_t(u) = \overline{P}_t(u) + \overline{P}_t(u + 2) = \sum_{k=-\infty}^{\infty} p_t(u - 2k), \quad u \in [-1,\, 1], \qquad (6.4)$$

$\pi_t(u) = 0$, $u \notin [-1,\, 1]$. Remark that the formula (6.4) is true for any symmetric function $p_t(u)$.

Suppose that r.v. z is distributed on $[-1,\, 1]$, its density $\mu(v)$ has the continuous derivative,
$$\frac{d\mu(v)}{dv} = 0, \quad v = \pm 1,$$
and r.v. z does not depend on the random process $y(t)$. Then the formula (6.3) allows to calculate the distribution density $\pi_t^z(u)$ of the random process $Z(t) = g(f(y(t) - 1 + z))$, $t \geq 0$, at the moment $t > 0$:

$$\pi_t^z(u) = \int_{-1}^{1} (\overline{P}_t(u - v) + \overline{P}_t(u + 2 + v))\mu(v)dv. \qquad (6.5)$$

Remark that the formulas (6.4), (6.5), which give the distribution of the diffusion reflected process, are analogous to the reflection method formulas which give the solution of the wave equation for a finite string with fixed edges [94, the chapt. III, §13, the points 5, 6].

As f is the homomorphism of the additive group R onto the group C, then for $0 < t_1 < t_2$

$$f(y(t_2)) = f(y(t_1)) \oplus f(y(t_2) - y(t_1)).$$

So the random process $f(y(t))$, $t \geq 0$, with independent and stable distributed increments is the homogenous Markov process with the state set $[-2,\ 2)$ and the symmetric distribution density of the transition from the state u into the state v during the time t

$$q_t(u,v) = \overline{P}_t(v \ominus u) = \overline{P}_t(v - u), \quad -2 \leq u,\ v < 2.$$

Denote $Q_t = \inf\{q_t(u,v),\ -2 \leq u,\ v < 2\}$ then from (6.3) obtain

$$1 = \int_{-2}^{2} \overline{P}_t(v - u)dv = \int_{-2}^{2} \overline{P}_t(v)dv \geq 4 \inf_{-2 \leq v < 2} \overline{P}_t(v)dv =$$

$$= 4Q_t \geq 4 \inf_{-2 \leq v < 2} p_t(v) > 0$$

and consequently

$$0 < Q_t < \frac{1}{4}, \quad t > 0. \tag{6.6}$$

Suppose that

$$P = P(u) = \begin{cases} 1/4, & u \in [-2,\ 2), \\ 0, & u \notin [-2,\ 2), \end{cases} \tag{6.7}$$

$$\pi = \pi(u) = \begin{cases} 1/2, & u \in [-1,\ 1], \\ 0, & u \notin [-1,\ 1]. \end{cases}$$

For the functions φ, ψ on $(-\infty,\ \infty)$ define the norm

$$||\varphi - \psi|| = \sup\{|\varphi(u) - \psi(u)|,\ -\infty < u < \infty\}$$

and denote by $[z]$ the integer part of the real number z.

Lemma 6.1. *For arbitrary* $h > 0,\ t \geq h$

$$||\pi_t - \pi|| \leq 2(1 - 4Q_h)^{k-1}||P_h - P||, \quad k = [t/h]. \tag{6.8}$$

Proof. As the transition density $q_t(u, v)$ is symmetric then the formula (6.7) gives

$$\int_{-2}^{2} P(v)q_t(v, u)dv = \frac{1}{4}\int_{-2}^{2} q_t(u, v)dv = P(u), \quad -2 \leq u < 2. \tag{6.9}$$

If $\Delta q_t = q_t - Q_t \geq 0$ then the formulas (6.6), (6.9) give for $-2 \leq u < 2$

$$||P_{t+\tau}(u) - P(u)|| =$$

$$= \sup\left(\left|\int_{-2}^{2} P_t(v)q_\tau(v, u)dv - \int_{-2}^{2} P(v)q_\tau(v, u)dv\right|, \; -2 \leq u < 2\right) =$$

$$= \sup\left(\left|\int_{-2}^{2} P_t(v)\Delta q_\tau(v, u)dv - \int_{-2}^{2} P(v)\Delta q_\tau(v, u)dv\right|, \; -2 \leq u < 2\right),$$

and for $t, \tau \geq 0$

$$||P_{t+\tau} - P|| \leq ||P_t - P|| \sup\left(\int_{-2}^{2} \Delta q_\tau(v, u)dv, \; -2 \leq u < 2\right) =$$

$$= ||P_t - P||(1 - 4Q_\tau). \tag{6.10}$$

The formulas (6.9), (6.10) allow to obtain by the induction:

$$||P_k - P|| \leq (1 - 4Q_1)^{k-1}||P_1 - P||, \quad k \geq 1. \tag{6.11}$$

Using the formulas (6.6), (6.10) it is easy to generalize the inequality (6.11) onto arbitrary $t \geq 1$

$$||P_t - P|| \leq (1 - 4Q_1)^{k-1}||P_1 - P||, \quad k = [t]. \tag{6.12}$$

The formulas (6.4), (6.12) give

$$||\pi_t - \pi|| \leq 2||P_t - P|| \leq 2(1 - 4Q_1)^{k-1}||P_1 - P||, \tag{6.13}$$

$$k = [t], \; t \geq 1.$$

The inequality (6.13) may be rewritten for the arbitrary positive number h in the form (6.8).

Remark that the lemma 6.1 is true for any symmetrical distribution density $p_t(u)$, which satisfies the condition (6.6).

Diffusion on Interval $[-r, r]$

Suppose that $m > 0$, $r = m^{-1/a}$, and consider the Markov process

$$Y_r(t) = rg(f(y(t)/r)), \; t \geq 0.$$

The process $Y_r(t)$, $t \geq 0$, is obtained from the process $y(t)$, $t \geq 0$, by the reflections from the edges of the interval $[-r, \; r]$. Denote by $\pi_{t,r} = \pi_{t,r}(u)$ the distribution density of r.v. $Y_r(t)$. It is obvious that $\pi_{t,1} = \pi_t$. Introduce normalized r.v.

$$W_{t,r} = \frac{Y_r(t)}{r} = g(f(y(t)/r)), \; t \geq 0,$$

with the distribution density $\psi_{t,r}(u) = r\pi_{t,r}(ru)$.

Lemma 6.2. *For* $K = [tm]$, $tm \geq 1$,

$$||\pi_{t,r}(u) - \pi(u/r)/r|| \leq \frac{2(1 - 4Q_1)^{K-1}||P_1 - P||}{r}. \tag{6.14}$$

Proof. The definition of the stable distribution (see [24, the chapter 17, §5]) gives that for $t > 0$ r.v.'s $y(tm)$, $y(t)/r$ coincide by the distribution. So r.v.'s $W_{t,r}$, $g(f(y(tm)))$ coincide by the distribution too and from (6.13)

$$||\psi_{t,r} - \pi|| = ||\pi_{tm} - \pi|| \leq 2(1 - 4Q_1)^{K-1}||P_1 - P||, \quad K = [tm], \quad tm \geq 1.$$

The formula (6.14) is proved.

Remark 3. *The formula (6.14) may be interpreted as an increasing of anomalous diffusion characteristic time on the interval $[-r, r]$ (or a decreasing dependently on r) in r^a times in a comparison with the interval $[-1, 1]$.*

Remark 4. *For $r < 1$ the anomalous diffusion "works" more slow and for $r > 1$ - more fast than the normal diffusion.*

Numerical experiment

Here the analytical results of the last paragraphs are compared with the results of a numerical experiment. In this experiment a closeness of the probability densities π_t, π is investigated for different $t > 0$. For this aim we simulate independent r.v.'s, which coincide with r.v. $Y(t)$ and so with r.v. $g(f(t^{1/a}y(1)))$ by the distribution. R.v. $y(1)$ is simulated approximately by the normalized sum

$$\widehat{y}(1) = \frac{v_1 + \ldots + v_N}{N^{1/a}}$$

of i.i.d.r.v.'s $v_1, \ldots, v_N$,

$$P(v_1 > t) = \frac{t^{-a}}{2}, \quad P(v_1 < -t) = \frac{|t|^{-a}}{2}, \quad t > 1.$$

Using M independent simulations of r.v. $g(f(t^{1/a}\hat{y}(1)))$ it is possible to calculate the frequencies $S_j(t)$, $j = 0, \ldots, 9$ of these simulations scorings into the sets

$$\left[-1 + \frac{2j}{10}, \ -1 + \frac{2(j+1)}{10}\right), \ j = 0, \ldots, 8, \quad \left[1 - \frac{2j}{10}, \ 1\right], \ j = 9.$$

We calculate the quantities

$$S(t) = 10M \sum_{j=0}^{9} \left(S_j(t) - \frac{1}{10}\right)^2$$

which characterize (analogously to χ^2-statistics) the deviations of $Y(t)$ distribution densities from the uniform density for different $t > 0$.

t	$S(t)$ for $a{=}1.9$	$S(t)$ for $a{=}1.95$	$S(t)$ for $a{=}1.99$
0,01	9169.31	7867.49	6584.26
0,02	4055.39	3053.09	2537.75
0,03	2200.99	1298.15	1017.17
0,04	1061.13	609.86	347.35
0,05	488.14	289.15	185.43
0,06	220.69	177.32	122.32
0,07	135.03	52.97	39.26
0,08	102.79	39.37	13.26
0,09	54.89	14.55	12.78

Table 2

Results of these calculations, represented in the table 2, show how fast do the distributions of $Y(t)$ converge to the uniform distribution if the time t increases. A qualitative coincidence of the numerical experiment results with the formula (6.8) is demonstrated.

Diffusion with periodical initial conditions

One-dimension case. To model this process take the natural number n and define the Markov process

$$Z_n(t) = g(f(y(t) - 1 + z_n)), \ t \geq 0,$$

where r.v. z_n has the uniform distribution on the finite set

$$I_n = \left\{ s = \frac{2k+1}{n} : \ k = 0, 1, \ldots, n-1 \right\}$$

and z_n, $y(t)$ are independent. Then the random process $Z_n(t)$, $t \geq 0$ may be considered as the anomalous diffusion on the interval $[-1, \ 1]$ but with periodical initial conditions

$$P(Z_n(0) = -1 + s) = \frac{1}{n}, \ \ s \in I_n. \tag{6.15}$$

Denote by $\Pi_{t,\,n} = \Pi_{t,\,n}(u)$, $\ \ P_{t,\,n} = P_{t,\,n}(u)$ the distribution densities of r.v.'s $Z_n(t)$, $f(y(t) - 1 + z_n)$, $\ t > 0$.

Lemma 6.3. *The following formula is true*

$$\Pi_{t,\,n}(u) = \frac{1}{n} \sum_{k=0}^{n-1} \pi_{t,\,1/n}\left(u + 1 - \frac{2k+1}{n} \right). \tag{6.16}$$

Proof. Analogously to the formulas $(6.3) - (6.5)$ obtain

$$P_{t,\,n}(u) = \frac{1}{n} \sum_{s \in I_n} P_t(u + 1 - s) = \frac{1}{n} \sum_{s \in I_n} \sum_{k=-\infty}^{\infty} p_t(u - 4k + 1 - s),$$

$$-2 \leq u < 2,$$

$$\Pi_{t,\,n}(u) = \sum_{v:\, g(v)=u} P_{t,\,n}(v) = \frac{1}{n} \sum_{s \in I_n} (\overline{P}_t(u + 1 - s) + \overline{P}_t(u + 1 + s)),$$

$$-1 \leq u \leq 1.$$

Then

$$\Pi_{t,\,n}(u) = \frac{1}{n} \sum_{k=-\infty}^{\infty} p_t\left(u - \frac{2k}{n} \right), \ \ -1 \leq u \leq 1. \tag{6.17}$$

The formula (6.17) leads to

$$\Pi_{t,\,n}(u) = \Pi_{t,\,n}\left(u + \frac{2}{n} \right), \ \ -1 \leq u \leq 1 - \frac{2}{n}. \tag{6.18}$$

Calculate now the function

$$\Pi_{t,\,n}(u), \quad -1 \le u < -1 + \frac{2}{n}.$$

For this aim take

$$u = w - 1 + \frac{1}{n}, \quad -\frac{1}{n} \le w < \frac{1}{n}.$$

Then in accordingly to (6.4), (6.17)

$$\Pi_{t,\,n}(w) = \frac{\pi_{t,\,1/n}(w)}{n}, \quad -\frac{1}{n} \le w < \frac{1}{n}. \tag{6.19}$$

The formulas (6.18), (6.19) lead to the equality (6.16).

The equality (6.16) means that the diffusion (normal or anomalous) on the set $[-1,\,1]$ with the periodical initial conditions (6.15) and the reflecting edges leads to the same result as the diffusion on the isolated (by the reflecting edges) subsets

$$\left[-1 + \frac{2k+1}{n} - \frac{1}{n}, \; -1 + \frac{2k+3}{n} + \frac{1}{n} \right], \quad k = 0, \ldots, n-1$$

of the set $[-1,\,1]$. Remark that the equality is true for each process $y(t)$ with the independent and symmetrically distributed increments.

Theorem 6.1. *For $tn^a \ge 1, \; L = [tn^a]$*

$$||\Pi_{t,\,n} - \pi|| \le 2(1 - 4Q_1)^{L-1}||P_1 - P||. \tag{6.20}$$

Proof. The statement of the theorem 6.1 is obtained directly from the equality (6.16) and the formula (6.14) in which r equals to $1/n$. $\qquad\square$

The inequality (6.20) may be interpreted as a decreasing of the characteristic time into n^a times in the model of the anomalous diffusion with n–periodical initial conditions (6.15). This result may be easily spread onto general case when r.v. $Z_n(0)$ has the distribution density with the continuous derivative $r_n(u)$, which satisfies the periodicity conditions:

$$r_n(u) = r_n\left(u + \frac{2}{n}\right), \quad -1 \le u \le 1 - \frac{2}{n},$$

and the symmetry conditions

$$r_n(-1 + \frac{1}{n} - v) = r_n\left(-1 + \frac{1}{n} + v\right), \quad 0 \le v \le \frac{1}{n},$$

and the boundary condition

$$\frac{dr_n(v)}{dv} = 0, \quad v = \pm 1.$$

It is a generalization of [7] results from the normal onto the anomalous diffusion.

Remark 5. *Introduce on the set* $[-1, 1)$ *the binary operation* " $\boxplus$ " *and the unary operation of the inverse* " $\boxminus$ " :

$$u \boxplus v = f^1(u + v), \quad \boxminus u = f^1(-u), \quad u, \ v \in [-1, 1)$$

with $f^1(u) = (u+1)/\mathrm{mod}\, 2 - 1$. *The group with the operations* "$\boxplus$", "$\boxminus$" *is denoted by* C^1. *It is easy to check that the transformation* $f^1 : (R, \ +) \rightarrow (C^1, \ \boxplus)$ *is the homomorphism of the additive group* R *of the real numbers onto the group* C^1.

Denote $P_t^1(u)$, $P_{t,\,n}^1(u)$ the probability densities of r.v.'s $f^1(y(t))$, $f^1(y(t) - 1 + z_n)$, $t > 0$, then for $u \in [-1, 1)$

$$P_t^1(u) = \sum_{k=-\infty}^{\infty} p_t(u - 2k),$$

$$P_{t,n}^1(u) = \frac{1}{n} \sum_{k=0}^{n-1} P_t^1\left(u + 1 - \frac{2k+1}{n}\right) = \frac{1}{n} \sum_{k=-\infty}^{\infty} p_t\left(u - \frac{2k+1}{n}\right).$$

From the equality (6.17) obtain

$$\Pi_{t,n}(u) = P_t^1(u), \ u \in [-1, 1). \tag{6.21}$$

The function $P_t^1(u)$, $u \in [-1, 1)$, characterizes the distribution of the diffusion process on the group C^1 at the moment t. This group may be considered as the circle with the length 2. The equality (6.21) allows to spread the theorem 6.1 results (concerning the diffusion on the interval with the reflecting edges and the periodical initial conditions) onto the diffusion on the circle with the length 2 and with the periodical initial conditions.

Multi-dimension anomalous diffusion. Results of the last paragraph may be generalized onto two-dimensional case if the interval $[-1, 1]$ is replaced by the square $[-1, 1] \times [-1, 1]$ and the process of the anomalous diffusion on the square with the reflecting boundaries is considered as two independent processes of the one dimension anomalous diffusion on the intervals $[-1, 1]$ with the reflecting edges. Disintegrate the unit square by the rectangular network with n^2 equal squares. Suppose that he initial state of the two-dimensional diffusion process on the unit square has the uniform distribution on the set of these n^2 squares centers. Then all one-dimensional estimates of the convergence rate are spread onto the two-dimensional case.

Denote all the one-dimensional distributions from the last paragraph by the indexes, characterized the numbers of the appropriate coordinates $j = 1, 2$:

$$P^{(j)} = P^{(j)}(u_j), \quad P_1^{(j)} = P_1^{(j)}(u_j), \quad \pi^{(j)} = \pi^{(j)}(u_j), \quad \Pi_{t,\,n}^{(j)} = \Pi_{t,\,n}^{(j)}(u_j).$$

Accordingly to the model of the two-dimensional diffusion (as the pair of the independent one-dimensional diffusion models) obtain:

$$\pi = \pi(u_1,\, u_2) = \pi^{(1)}(u_1)\pi^{(2)}(u_2),$$

$$\Pi_{t,\,n} = \Pi_{t,\,n}(u_1,\, u_2) = \Pi_{t,\,n}^{(1)}(u_1)\Pi_{t,\,n}^{(2)}(u_2).$$

From the inequality (6.20) for $L = [tn^{1/a}], \quad tn^{1/a} \geq 1, \quad j = 1, 2$

$$||\Pi_{t,\,n}^{(j)} - \pi^{(j)}|| \leq 2(1 - 4Q_1)^{L-1}||P_1^{(j)} - P^{(j)}|| = \Delta_{t,\,n}^{(j)}. \tag{6.22}$$

Define the norm $||\Phi - \Psi|| = \sup\{|\Phi(u_1,\, u_2) - \Psi(u_1,\, u_2)|, \; -\infty < u_1,\, u_2 < \infty\}$ of the functions Φ, Ψ on the plane and put $\Delta_{t,\,n} =$
$= \Delta_{t,\,n}^{(1)} - \Delta_{t,\,n}^{(2)}.$

Theorem 6.2. *For $L = [tn^{1/a}], \quad tn^{1/a} \geq 1,$*

$$||\Pi_{t,\,n} - \pi|| \leq \Delta_{t,\,n}(1 + \Delta_{t,\,n}). \tag{6.23}$$

Proof. From the triangle inequality obtain

$$||\Pi_{t,\,n} - \pi|| = ||\Pi_{t,\,n}^{(1)}\Pi_{t,\,n}^{(2)} - \pi^{(1)}\pi^{(2)}|| \leq ||\Pi_{t,\,n}^{(1)}\Pi_{t,\,n}^{(2)} - \pi^{(1)}\Pi_{t,\,n}^{(2)}|| +$$

$$+ ||\pi^{(1)}\Pi_{t,\,n}^{(2)} - \pi^{(1)}\pi^{(2)}|| \leq ||\Pi_{t,\,n}^{(2)}|| \, ||\Pi_{t,\,n}^{(1)} - \pi^{(1)}|| + ||\pi^{(1)}|| \, ||\Pi_{t,\,n}^{(2)} - \pi^{(2)}|| \leq$$

$$\leq (||\pi^{(2)}|| + ||\pi^{(2)} - \Pi_{t,\,n}^{(2)}||) \, ||\Pi_{t,\,n}^{(1)} - \pi^{(1)}|| + ||\pi^{(1)}|| \, ||\Pi_{t,\,n}^{(2)} - \pi^{(2)}|| \leq$$

$$\leq \left(\frac{1}{2} + ||\pi^{(2)} - \Pi_{t,\,n}^{(2)}||\right) ||\Pi_{t,\,n}^{(1)} - \pi^{(1)}|| + \frac{1}{2} ||\Pi_{t,\,n}^{(2)} - \pi^{(2)}||.$$

Using the formula (6.22) obtain the inequality (6.23).

As $\Delta_{t,\,n} \to 0,\ \ t \to \infty$ so the convergence rate of the two-dimensional case (6.23) is analogous to the one-dimensional case (6.20). This result may be easily spread onto the multi-dimensional case. A specifics of this multi-dimensional diffusion model is that as one-dimensional diffusion components so their initial conditions are independent.

The equality (6.21) allows to spread the theorem 6.2, analyzed the diffusion on the square (with the reflecting sides and the periodical initial conditions) onto the diffusion on the torus with the side length 2 and with periodical initial conditions.

Multi-dimension normal diffusion. Consider k-dimension normal diffusion model in which $\Pi_{t,\,n}(y_1, \ldots, y_k)$ is the distribution density at the moment t in the point $(y_1, \ldots, y_k) \in [-1, 1]^k$:

$$
\left(\frac{\partial}{\partial t} - \sum_{j=1}^{k} \frac{\partial^2}{\partial y_j^2} \right) \Pi_{t,\,n}(y_1, \ldots, y_k) = 0,
$$

$$
\frac{\partial}{\partial y_j} \Pi_{t,\,n}(\pm 1, \ldots, \pm 1, y_j, \pm 1, \ldots, \pm 1) = 0, \quad -1 \le y_j \le 1, \quad j = 1, \ldots, k,
$$

$$
\Pi_{0,\,n}(y_1, \ldots, y_k) = \frac{1}{2^k} + \sum_{j_1,\ldots,j_k=1}^{m} a(j_1, \ldots, j_k) \prod_{r=1}^{k} \cos \pi n j_r y_r > 0, \quad m < \infty.
$$

Then

$$
\Pi_{t,\,n}(y_1, \ldots, y_k) =
$$

$$
= \frac{1}{2^k} + \sum_{j_1,\ldots,j_k=1}^{m} a(j_1, \ldots, j_k) \prod_{r=1}^{k} \cos \pi n j_r y_r \exp\left(-\pi^2 n^2 t \sum_{j=1}^{m} k_j^2 \right)
$$

and for $|a(1, \ldots, 1)| > 0$

$$
\left\| \Pi_{t,\,n}(y_1, \ldots, y_k) - \frac{1}{2^k} \right\| \sim |a(1, \ldots, 1)| \exp(-\pi^2 k n^2 t), \quad t \to \infty.
$$

So it is possible to choose n-periodical (by each coordinate) initial conditions so that in the case of the normal diffusion on $k-$dimension square $[-1, 1]^k$ the characteristic time of the diffusion decreases into kn^2 times.

6.2. Random time of threads wisp break

A problem of threads wisp break analysis is model for an investigation of composites solidity [61, p. 715–721]. A base of this problem solution is the dependence of the probability $P(L, \delta)$ that the thread with the length L is not broken under the stress δ :

$$P(L, \delta) = \exp\left(-\frac{L}{L_0}\left(\frac{\delta}{\delta_*}\right)^{\alpha}\right). \tag{6.24}$$

Here L_0 is characteristic unit, δ_* is characteristic stress. The dependence was found experimentally by Freidental.

In [89] the probability of the break of two threads wisp with across connections is investigated. Quantitative estimate of the wisp strengthening by across connections was made. But a spread of this result onto arbitrary number of threads leads to large difficulties as in a construction of stochastic model so in its investigation. Main cause of these difficulties is in static character of this model (6.24). But the break of the wisp with arbitrary number of threads demands dynamical analysis. In this subsection such difficulties are got over.

Denote by $\tau(L, \delta)$ random time of the break in the thread with the length L under the stress δ. Suppose that the formula (6.24) characterizes the thread break with (conditional) unit length during (conditional) unit time:

$$P\{\tau(L, \delta) > 1\} = P(L, \delta). \tag{6.25}$$

Ignoring aging and after-action processes and choose for r.v. $\tau(L, \delta)$ exponential distribution which generalizes the formula (6.25):

$$P\{\tau(L, \delta) > t\} = \exp\left(-\frac{L}{L_0}\left(\frac{\delta}{\delta_*}\right)^{\alpha} t\right). \tag{6.26}$$

Using the formula (6.26) it is possible to construct stochastic process of threads wisp break.

Random time of threads wisp break

Consider n identical threads with the length L, connecting at the initial and final points A, B. Each thread is under the stress δ. Following the monograph [42] methods, describe the random process $x(t)$ of threads wisp break. Here $x(t)$ is the number of unbroken threads in the wisp at the moment $t \geq 0$, $x(0) = n$. Denote

$t_0 = 0$. $t_1 = t_0 + \tau_1$. $t_2 = t_1 + \tau_2, \ldots, t_n = t_{n-1} + \tau_n$ — the sequential broken moments of the considered wisp. For $t_{i-1} \leq t < t_i$ the equality $x(t) = n - i + 1$ is true and so the stress in each among $(n - i + 1)$ unbroken threads is

$$\delta_i = \frac{n\delta}{n - i + 1}, \quad i = 1, \ldots, n. \tag{6.27}$$

Using the formulas (6.26), (6.27) and supposing that the break processes in different threads are independent, find in the half-interval $t_{i-1} \leq t < t_i$:

$$P\{\tau_i > t\} = e^{-t/T_i}, \quad T_i = a\frac{(n - i + 1)^{\alpha-1}}{n^\alpha},$$
$$a = \frac{L_0}{L}\left(\frac{\delta_*}{\delta}\right)^\alpha, \quad i = 1, \ldots, n, \tag{6.28}$$

r.v.'s $\tau_1, \ldots, \tau_n$ are independent. Accordingly to the formula (6.28) and to the independence condition for r.v's $\tau_1, \ldots, \tau_n$ obtain that

$$\eta_1 = \frac{\tau_1}{T_1}, \ldots, \eta_n = \frac{\tau_n}{T_n}$$

are independent and identically distributed,

$$P\{\eta_i > t\} = e^{-t}, \quad i = 1, \ldots, n. \tag{6.29}$$

So the random moment t_n of the complete wisp break satisfies the equality

$$t_n = \sum_{i=1}^{n} T_i \eta_i. \tag{6.30}$$

Using the independence of r.v.'s $\eta_1, \ldots, \eta_n$ and the formulas (6.29), (6.30), obtain:

$$Mt_n = \sum_{i=1}^{n} T_i, \quad Dt_n = \sum_{i=1}^{n} T_i^2. \tag{6.31}$$

Put the formula (6.28) into the equalities (6.31) and find:

$$Mt_n = \frac{am_n(\alpha - 1)}{n^\alpha}, \quad Dt_n = \frac{a^2 m_n(2(\alpha - 1))}{n^{2\alpha}} \tag{6.32}$$

where $m_n(\gamma) = \sum_{i=1}^{n} i^\gamma$, $-\infty < \gamma < \infty$. Transiting from the summation to the integration, obtain the following inequalities:

$$n^\gamma/\gamma \leq m_n(\gamma - 1) \leq (n + 1)^\gamma/\gamma, \quad \gamma > 1,$$

$$(n^\gamma - 1)/\gamma \le m_n(\gamma - 1) \le n^\gamma/\gamma, \quad 0 < \gamma \le 1,$$

$$\ln(1 + n) \le m_n(\gamma - 1) \le 1 + \ln n, \quad \gamma = 0,$$

$$((n + 1)^\gamma - 1)/\gamma \le m_n(\gamma - 1) \le 1 + (n^\gamma - 1)/\gamma, \quad \gamma < 0. \tag{6.33}$$

Combine the formulas (6.32), (6.33) and find:

$$\left| Mt_n - \frac{a}{\alpha} \right| \le \begin{cases} 2^{\alpha-1}a/n, & 1 < \alpha, \\ a/(\alpha n^\alpha), & 0 < \alpha \le 1, \end{cases}$$

$$Dt_n \le \begin{cases} 2^{2\alpha-1}a^2/n, & \alpha > 1/2, \\ (1 + \ln n)a^2/n, & \alpha = 1/2, \\ 2(1 - \alpha)a^2/(n^{2\alpha}(1 - 2\alpha)), & 0 < \alpha < 1/2. \end{cases}$$

From the obvious inequality

$$M\left(t_n - \frac{a}{\alpha}\right)^2 \le \left(Mt_n - \frac{a}{\alpha}\right)^2 + Dt_n$$

finally obtain:

$$M\left(t_n - \frac{a}{\alpha}\right)^2 \le r_n \tag{6.34}$$

with

$$r_n = \begin{cases} \left(1/\alpha^2 + 2(1 - \alpha)/(1 - 2\alpha)\right) a^2/n^{2\alpha}, & 0 < \alpha < 1/2, \\ \left(1/\alpha^2 + 1 + \ln n\right) a^2/n, & \alpha = 1/2, \\ \left(2^{2\alpha-1} + 1/(\alpha^2 n^{2\alpha-1})\right) a^2/n, & 1/2 < \alpha \le 1, \\ \left(2^{2\alpha-1} + 2^{2(\alpha-1)}/n\right) a^2/n, & 1 < \alpha. \end{cases} \tag{6.35}$$

Accordingly to the Chebyshev inequality and the formula (6.34) for each $\varepsilon > 0$

$$P\left\{ \left| t_n - \frac{a}{\alpha} \right| > \varepsilon \right\} \le \frac{r_n}{\varepsilon^2}. \tag{6.36}$$

Connecting the formulas (6.35), (6.36), obtain that the random time t_n of n threads (with the length L, under the stress δ and with the same initial and final points) break tends by the probability to the constant a/α if $n \to \infty$.

Influence of across connections on random time
of threads wisp break

Suppose now that all n threads are divided into m equal parts by the points $A_0 = A,\ A_1, \ldots, A_{m-1},\ A_m = B$ and are connected in these points with each other. Define $t_{n,k}^{(m)}$ as the moment of the break between the points $A_{k-1},\ A_k,\ k = 1, \ldots, m$. Accordingly to the formula (6.26) it is naturally to consider that r.v.'s $t_{n,1}^{(m)}, \ldots, t_{n,k}^{(m)}$ are independent with common distribution which coincides with r.v. mt_n distribution. Then break time of the whole wisp between the points A, B satisfies the formula

$$t_n^{(m)} = \min\left(t_{n,1}^{(m)}, \ldots, t_{n,m}^{(m)}\right).\tag{6.37}$$

Define independent r.v.'s with the distributions, which coincide with r.v. t_n distribution:

$$t_{n,1} = \frac{t_{n,1}^{(m)}}{m}, \ldots, t_{n,m} = \frac{t_{n,1}^{(m)}}{m}.$$

Then from the formula (6.37) obtain

$$t_n^{(m)} = m \min(t_{n,1}, \ldots, t_{n,m}).\tag{6.38}$$

Using the formulas (6.34), (6.38), estimate the mean

$$M\left|t_n^{(m)} - \frac{ma}{\alpha}\right| = mM\left|\min(t_{n,1}, \ldots, t_{n,m}) - \frac{a}{\alpha}\right| \leq$$

$$\leq mM\sum_{i=1}^{m}\left|t_{n,1} - \frac{a}{\alpha}\right| = m^2 M\left|t_n - \frac{a}{\alpha}\right| \leq$$

$$\leq m^2\sqrt{M\left(t_n - \frac{a}{\alpha}\right)^2} = m^2\sqrt{r_n}.\tag{6.39}$$

The formula (6.39) leads to the inequalities

$$\frac{ma}{\alpha}\left(1 - \frac{\alpha}{a}m\sqrt{r_n}\right) \leq Mt_n^{(m)} \leq \frac{ma}{\alpha}\left(1 + \frac{\alpha}{a}m\sqrt{r_n}\right),$$

$$P\left\{\left|t_n^{(m)} - \frac{ma}{\alpha}\right| > \frac{ma}{\alpha}\varepsilon\right\} \leq \frac{\alpha m\sqrt{r_n}}{a\varepsilon}.\tag{6.40}$$

Suppose now that for some $q > 0$ the parameters m, n satisfy the equalities

$$m = \frac{n^q}{\ln n}, \quad \lim_{n \to \infty} m\sqrt{r_n} = 0.\tag{6.41}$$

Then accordingly to (6.40) obtain

$$\lim_{n\to\infty} Mt_n^{(m)}/(ma/\alpha) = 1.\tag{6.42}$$

R.v.'s $t_n^{(m)}/(ma/\alpha)$ tend by the probability to 1 if $n \to \infty$. The formulas (6.36). (6.42) comparison allows to say that the introduction of m across connections into the wisp leads to the increasing of its break time into m tines (a choice of the constant q in the formula (6.41) is made accordingly to the inequalities $0 \le q \le \min(1/2, \alpha)$.

Suppose now that n is fixed and $m \to \infty$. Then accordingly to the equality (6.38) and the limit theorem for the minimum of m i.i.d.r.v.'s [99] the distribution of r.v. $t_n^{(m)}$ tends to exponential d.f. The parameter of this d.f. does not depend on m. So the effect of m-divisible increasing of the break time disappears. The case when $m = n^q/\ln n$, $\min(1/2, \alpha) < q < \infty$, demands additional consideration.

6.3. Construction and investigation of stochastic model of sliding regime

An investigation of a motion in a vicinity of a sliding line attracts interest of specialists for a long time period. Deterministic models, using for this aim, allow to obtain different averaged sliding equations [93, p. 27]. But an estimate of sliding quality demands a knowledge of more detailed motion characteristics. A finding of these characteristics in frames of deterministic models is complicated in analytical calculations because the deterministic models origin pseudo-stochastic processes. In this subsection a stochastic model of sliding regime is constructed. This model allows to calculate as a deviation from sliding line so a non-uniformity of the motion along sliding line dependently on a sliding step (if it tends to zero).

Suppose that the motion in the vicinity of the sliding line $y = 0$ is considered in the moments kh, $k = 0, 1, \ldots$, $h > 0$, and is described by the following random sequence (x_k, y_k), $k = 0, 1, \ldots$ Denote

$$(\Delta x_k, \Delta y_k) = (x_{k+1} - x_k, y_{k+1} - y_k),$$

take some real numbers Δ^1, Δ^2 and positive numbers λ_1, λ_2, choose $x_0 = 0$, $y_0 = 0$ and put

$$(\Delta x_k, \Delta y_k) = \begin{cases} (\Delta^1, -\Delta_k^1), & y_k > 0, \\ (\Delta^2, \Delta_k^2), & y_k \le 0. \end{cases}\tag{6.43}$$

Here $\Delta_k^1, \Delta_k^2, \ k = 0, 1, \ldots$, are independent r.v.'s so that for $u \geq 0$

$$P\{\Delta_k^i > u\} = e^{-\lambda_i u}, \quad i = 1, 2, \quad k = 0, 1, \ldots. \tag{6.44}$$

Define the random processes $x^{(1)}(t)$, $y^{(1)}(t)$, $t \geq 0$, $x^{(1)}(0) = 0$, $y^{(1)}(0) = 0$, characterizing the motion along the sliding line and across it, by the equalities

$$x^{(1)}((k+\tau)h) = x_k, \quad 0 \leq \tau < 1, \quad k \geq 0,$$

$$\tag{6.45}$$

$$y^{(1)}((k+\tau)h) = y_k, \quad 0 \leq \tau < 1, \quad k \geq 0.$$

Suppose that the time step h depends on the parameter (of the step cutting) m: $h = 1/m$ and define the random processes

$$x^{(m)} = x^{(m)}(t) = x(mt)/m, \quad t \geq 0,$$

$$\tag{6.46}$$

$$y^{(m)} = y^{(m)}(t) = y(mt)/m, \quad t \geq 0.$$

The formulas (6.46) characterize m-divisible cutting of the sliding step.

Denote

$$p^1 = \frac{\lambda_1}{\lambda_1 + \lambda_2}, \ p^2 = \frac{\lambda_2}{\lambda_1 + \lambda_2}$$

and define accordingly to [88] the process of the average sliding by the equalities

$$\bar{x} = \bar{x}(t) = (p^1 \Delta^1 + p^2 \Delta^2)t, \ \bar{y} = \bar{y}(t) = 0. \tag{6.47}$$

Introduce on the set $U = \{u = u(t) : M \int_0^1 u^2(t)dt < \infty\}$ of the random processes the distance ρ by the equality

$$\rho(u_1, u_2) = \left(M \int_0^1 (u_1(t) - u_2(t))^2 dt \right)^{1/2}, \quad u_1, u_2 \in U,$$

and denote $\rho_m = \rho(x^{(m)}, \bar{x})$. The quantity ρ_m characterizes the non-uniformity of the sliding along the line $y = 0$.

Define now the maximal deviation of the motion trajectory from the sliding line at the unit time interval $0 \leq t \leq 1$:

$$Y_m = \sup_{0 \leq t \leq 1} \left| y^{(m)}(t) \right| = \frac{1}{m} \sup_{0 \leq t \leq 1} |y(mt)| = \frac{1}{m} \max_{1 \leq k \leq m} |y_k| \,. \tag{6.48}$$

Denote by

$$\kappa_m = \inf\{u : P\{Y_m \geq u\} \leq u\} = \kappa(0, Y_m)$$

the Qui Fan distance [99, p. 35] between 0 and r.v. Y_m. The quantity κ_m characterizes the random deviation of the motion trajectory from the sliding line.

Quantitative estimate of sliding non-uniformity

Denote by $p_k = p_k(u)$ the densities of r.v.'s y_k, $k = 0, 1, \ldots$, $-\infty < u < +\infty$, distributions. Put

$$r_1 = r_1(u) = \begin{cases} 0, & u > 0, \\ \lambda_1 e^{\lambda_1 u}, & u \leq 0, \end{cases}$$

$$r_2 = r_2(u) = \begin{cases} \lambda_2 e^{-\lambda_2 u}, & u > 0, \\ 0, & u \leq 0, \end{cases}$$

and denote by **(A)** the property of the after action absence of the exponential distributions (6.44). Accordingly to (6.43), (6.44) obtain the equality

$$p_1 = r_2. \tag{6.49}$$

Taking into an account the property **(A)** and the formulas (6.43), (6.44), (6.49), find

$$p_2 = p^1 r_1 + p^2 r_2. \tag{6.50}$$

Using the formula (6.50) and the property **(A)**, obtain

$$p_3 = r_1(p^1 p^1 + p^2 p^1) + r_2(p^2 p^2 + p^1 p^2) = p^1 r_1 + p^2 r_2 = p_2. \tag{6.51}$$

Analogously to the formulas (6.50), (6.51) obtain the following chain of the equalities

$$p_2 = p_3 = \ldots = p^1 r_1 + p^2 r_2. \tag{6.52}$$

Accordingly to the formulas (6.43), (6.44), (6.52) and with the property **(A)** we have

$$P\{y_{k+1} > 0/y_k > 0\} = P\{y_{k+1} > 0/y_k \leq 0\} = p^1,$$

$$\tag{6.53}$$

$$P\{y_{k+1} \leq 0/y_k > 0\} = P\{y_{k+1} \leq 0/y_k \leq 0\} = p^2,$$

$$k = 1, 2, \ldots$$

Define the function f on the real axis by the equality

$$f(u) = \begin{cases} 1, & u > 0, \\ -1, & u \le 0. \end{cases}$$

Accordingly to the formulas (6.53) the sequence $f(y_2), f(y_3), \ldots$ consists of independent r.v.'s with common d.f.:

$$P\{f(y_k) = 1\} = p^1, \quad P\{f(y_k) = -1\} = p^2, \quad k = 2, 3. \ldots \tag{6.54}$$

Consider the random sequence $\Delta x_k, \; k = 0, 1, \ldots$. It is obvious that almost surely

$$\Delta x_0 = \Delta^2, \quad \Delta x_1 = \Delta^1. \tag{6.55}$$

R.v.'s $\delta_k = \Delta x_k, \; k = 2, 3, \ldots$, accordingly to the formulas (6.43), (6.54) and to the independence of $f(y_k), \; k = 2, 3, \ldots$, are independent too and

$$P\{\delta_k = \Delta^i\} = p^i, \quad i = 1, 2, \quad k = 2, 3, \ldots \tag{6.56}$$

Suppose that r.v.'s δ_0, δ_1 have d.f. (6.56) and together with $\delta_2, \delta_3, \ldots$ create the sequence of independent r.v.'s.

The random sequence $x_k, \; k = 0, 1, \ldots$, accordingly to the formulas (6.55), (6.56) satisfies the equalities

$$\begin{aligned} x_0 = 0, \quad x_1 &= \Delta^2, \quad x_2 = \Delta^1 + \Delta^2, \\ x_3 = \Delta^1 + \Delta^2 + \delta_2, \quad x_4 &= \Delta^1 + \Delta^2 + \delta_2 + \delta_3, \ldots. \end{aligned} \tag{6.57}$$

Define the random sequence

$$z_k = \sum_{j=0}^{k-1} \delta_j, \quad k = 0, 1, \ldots, \tag{6.58}$$

and the deterministic sequence

$$w_k = kM\delta_0, \quad k = 0, 1, \ldots \tag{6.59}$$

Using the sequences (6.58), (6.59), define the processes $z^{(m)} = z^{(m)}(t), \; w^{(m)} = w^{(m)}(t), \; t \ge 0$:

$$z^{(1)}((k + \tau)h) = z_k, \quad w^{(1)}((k + \tau)h) = w_k, \quad 0 \le \tau < 1, \quad k = 0, 1, \ldots,$$

$$z^{(m)}(t) = \frac{1}{m}z^{(1)}(mt), \quad w^{(m)}(t) = \frac{1}{m}w^{(1)}(mt).$$

Calculate the mean square distances

$$\rho'_m = \rho\left(x^{(m)}, z^{(m)}\right) = \frac{1}{m}\left(b + \frac{a-2b}{m}\right)^{1/2}, \tag{6.60}$$

$$\rho''_m = \rho\left(z^{(m)}, w^{(m)}\right) = \frac{1}{m^{1/2}}\left(\frac{c}{2}\left(1-\frac{1}{m}\right)\right)^{1/2}, \tag{6.61}$$

$$\rho'''_m = \rho\left(w^{(m)}, \bar{x}\right) = \frac{d}{m\sqrt{3}} \tag{6.62}$$

where $a = M(\delta_0 - \Delta^2)^2$, $b = M(\delta_0 - \Delta^2 + \delta_1 - \Delta^1)^2$, $c = D\delta_0$, $d = |M\delta_0|$. From the triangle inequality obtain

$$\rho''_m - \rho'_m - \rho'''_m \le \rho_m \le \rho''_m + \rho'_m + \rho'''_m. \tag{6.63}$$

Put the equalities (6.60), (6.61), (6.62) into the inequality (6.63):

$$\frac{1}{\sqrt{m}}\left[\left(\frac{c}{2}\left(1-\frac{1}{m}\right)\right)^{\frac{1}{2}} - \frac{1}{\sqrt{m}}\left(\left(b+\frac{a-2b}{m}\right)^{\frac{1}{2}} + \frac{d}{\sqrt{3}}\right)\right] \le$$

$$\le \rho_m \le \frac{1}{\sqrt{m}}\left[\left(\frac{c}{2}\left(1-\frac{1}{m}\right)\right)^{\frac{1}{2}} + \frac{1}{\sqrt{m}}\left(\left(b+\frac{a-2b}{m}\right)^{\frac{1}{2}} + \frac{d}{\sqrt{3}}\right)\right]. \tag{6.64}$$

So the following statement is proved.

Theorem 6.3. *If the conditions (6.43), (6.44) are true then the formula (6.64) and consequently the formula*

$$\rho_m \sim \sqrt{\frac{c}{2m}}, \quad m \to \infty \tag{6.65}$$

take place.

Quantitative estimate of maximal deviation of sliding regime trajectory from sliding line

Denote

$$n_0 = 0,$$

$$n_{2j+1} = \min(k : k > n_{2j}, \; y_k > 0), \tag{6.66}$$

$$n_{2j+2} = \min(k : k > n_{2j+1}, \; y_k \le 0), \; j \ge 0$$

and designate

$$Z_i = |y_{n_i}|, \quad i = 0, 1, \ldots.$$

From the formulas (6.43), (6.44), (6.66) and the property **(A)** obtain that r.v.'s Z_i, $i = 1, 2, \ldots$, are independent and

$$P\{Z_{2j+1} > u\} = e^{-\lambda_2 u}, \quad P\{Z_{2j+2} > u\} = e^{-\lambda_1 u}, \quad j = 0, 1, \ldots \tag{6.67}$$

From the formulas (6.48), (6.66) obtain that almost surely

$$\frac{Z_1}{m} \leq Y_m \leq \frac{\max(Z_n, 1 \leq i \leq m)}{m} = \frac{\max(A_m, B_m)}{m} \leq \frac{A_m}{m} + \frac{B_m}{m}, \tag{6.68}$$

$$A_m = \max(Z_{2j+1}, \ 1 \leq 2j + 1 \leq m),$$
$$B_m = \max(Z_{2j+2}, \ 1 \leq 2j + 2 \leq m).$$

The formulas (6.68) and the triangle inequality lead to

$$\kappa\left(0, \frac{Z_1}{m}\right) \leq \kappa_m \leq \kappa\left(0, \frac{A_m}{m} + \frac{B_m}{m}\right) \leq \kappa\left(0, \frac{A_m}{m}\right) +$$

$$+ \kappa\left(\frac{A_m}{m}, \frac{A_m}{m} + \frac{Z_m}{m}\right) = \kappa\left(0, \frac{A_m}{m}\right) + \kappa\left(0, \frac{B_m}{m}\right). \tag{6.69}$$

Calculate for

$$u_m = \frac{\ln m - \ln\left(\frac{\ln m}{\lambda_2}\right)}{\lambda_2 m}, \quad \frac{\ln m}{\lambda_2} > 1 \tag{6.70}$$

the probability

$$P\left\{\frac{Z_1}{m} > u_m\right\} = \frac{\ln m}{\lambda_2 m} > u_m. \tag{6.71}$$

The formula (6.71) leads to the inequality

$$u_m \leq \kappa\left(0, \frac{Z_1}{m}\right) \leq \kappa_m. \tag{6.72}$$

Estimate now the distances $\kappa(0, A_m/m)$, $\kappa(0, B_m/m)$. Is is obvious that for all $u \geq 0$ the following inequality

$$1 - \left(1 - e^{-\lambda m u}\right)^{m/2} \leq$$

$$\leq 1 - \exp\left(-\frac{m}{2} e^{-\lambda m u} / \left(1 - e^{-\lambda m u}\right)\right) = F_m(u) \tag{6.73}$$

is true. Choose $v_m = 2\ln m / (\lambda m)$ and calculate

$$F_m(v_m) = 1 - \exp\left(-\frac{1}{2m} \bigg/ \left(1 - \frac{1}{m^2}\right)\right) \le \frac{1}{2m} \bigg/ \left(1 - \frac{1}{m^2}\right). \qquad (6.74)$$

Consequently if the condition

$$\frac{2\ln m}{\lambda} > \frac{1}{2(1 - 1/m^2)} \qquad (6.75)$$

is true then

$$F_m(v_m) \le v_m. \qquad (6.76)$$

Suppose that in the formulas (6.73), (6.74), (6.75) $\lambda = \lambda_2$, $\lambda = \lambda_1$ then obtain from the formulas (6.68), (6.76):

$$\kappa\left(0, \frac{A_m}{m}\right) \le \frac{2\ln m}{\lambda_2 m}, \quad \kappa\left(0, \frac{B_m}{m}\right) \le \frac{2\ln m}{\lambda_1 m}. \qquad (6.77)$$

Add the inequalities (6.77) and obtain

$$\kappa\left(0, \frac{A_m}{m}\right) + \kappa\left(0, \frac{B_m}{m}\right) \le 2\left(\frac{1}{\lambda_1} + \frac{1}{\lambda_2}\right)\frac{\ln m}{m}. \qquad (6.78)$$

So the formulas (6.68) – (6.78) lead to the statement.

Theorem 6.4. *If the conditions*

$$\ln m > \max\left(\lambda_2, \frac{\lambda_1}{4(1 - 1/m^2)}, \frac{\lambda_2}{4(1 - 1/m^2)}\right) \qquad (6.79)$$

are true then:

$$\frac{\ln m}{\lambda_2 m}\left(1 - \frac{\ln\left(\frac{\ln m}{\lambda_2}\right)}{\ln m}\right) \le \kappa_m \le 2\left(\frac{1}{\lambda_1} + \frac{1}{\lambda_2}\right)\frac{\ln m}{m}. \qquad (6.80)$$

Accordingly to the theorem 6.3 (see the formula (6.65)) the quantity ρ_m, characterizing the sliding non-uniformity, decreases like $m^{-1/2}$. The theorem 6.4 (see the formulas (6.79), (6.80)) shows that for $m \to \infty$ the quantity κ_m, characterizing the maximal deviation of the sliding trajectory from the sliding line at the unit time interval, has the order $\ln m / m$. These results have been publicized in [90, 91].

Chapter 7

Cooperative and nonlinear effects in models of mathematical economics

In this chapter, we consider models of mathematical economics with cooperative and nonlinear effects. At first some linear stochastic growth models with a movement of capital between them (dependent on their current growth coefficients) is analyzed. Estimates of mean capital growth as a function of its movement frequency is constructed.

Then a modelling and investigation of debt discharge procedure is made. The main idea of the discharge is in its transformation into some auxiliary transportation problem. The analogous situation appeared in chapter 1 in the construction of the stochastic control of the Markov process parameter. Using this fact, we investigate a few properties of the discharge procedure: the influence of initial data and its errors on final results, debt discharge in isolated coalitions and by agents, etc.

At last, the Estes model of the self-learning is considered. This model is one of the first mathematical models in psychology. Recent investigations confirm that psychological factors play an important role in an economic decisions. Using the methods of the previous chapters it is possible to find new properties and a clear interpretation of the Estes model.

7.1. Cooperative effects in simplest stochastic growth model

Consider N manufactories with the unit mean rate growth (of an income or a production volume). Suppose that these rates have identical means and independent and identically distributed random disturbances. At an initial moment we put our money in some manufactory and then move the money to another manufactory with the largest current rate growth. The problem is to analyze how does the mean rate growth of our money depend on the motion frequency. This problem will be considered in frames of the following model.

Suppose that the capital $X_i(t)$ of the i-th manufactory is described by the following stochastic differential equation: $i = 1, \ldots, N$,

$$dX_i(t) = X_i(t)(\mu dt + \sigma dW_i(t)), \ X_i(0) = X_i^0 > 0. \tag{7.1}$$

Here $W_i(t)$, $i = 1, \ldots, N$, is the sequence of the independent and standard Wiener processes, $W_i(0) = 0$, $DW_i(t) = t$, and σ, μ are the constants, satisfying the conditions $\sigma > 0$, $\mu > 0$. In the model (7.1) the random growth rate of the i-th manufactory capital on the half interval $[t, t + dt)$ is following

$$\mu dt + \sigma(W_i(t + dt) - W_i(t))$$

with the mean μdt and with the random difference $\sigma(W_i(t+dt) - W_i(t))$. It is obvious that the equalities

$$X_i(t + \Delta t) = X_i(t) \exp(\mu \Delta t + \sigma \Delta W_i(t)),$$

$$\Delta W_i(t) = W_i(t + \Delta t) - W_i(t), \ \ t, \Delta t > 0, \ \ i = \overline{1, N}. \tag{7.2}$$

are true.

Describe now the motion of our money between these manufactories. Suppose that we put our money on the half interval $[t, t + \Delta t)$ into the manufactory with the maximal rate growth. That is the dynamics of our money satisfies the equality

$$X(t + \Delta t) = X(t) \exp(\mu \Delta t + \sigma \max_{1 \leq i \leq N} \Delta W_i(t)). \tag{7.3}$$

Suppose that the moments t_k of our money transmission satisfy the condition

$$t_k = \frac{k}{n}, \ \ k = 0, 1, \ldots. \tag{7.4}$$

Then accordingly to the formulas (7.3), (7.4) we have for $k > 0$:

$$X(t_k) = X(0) \prod_{s=0}^{k-1} \exp\left(\frac{\mu}{n} + \sigma \max_{1 \le i \le n}\left(W_i\left(\frac{s+1}{n}\right) - W_i\left(\frac{s}{n}\right)\right)\right). \tag{7.5}$$

The mean rate growth of our money is characterized by the quantity

$$A(n, N) = \frac{MX(1)}{X(0)}.$$

Using the formula (7.5), it is easy to obtain that

$$A(n, N) = e^\mu \left[M \exp\left(\sigma \max_{1 \le i \le N} W_i\left(\frac{1}{n}\right)\right)\right]^n =$$

$$= e^\mu \left[M \exp\left(\frac{\sigma}{\sqrt{n}} \max_{1 \le i \le N} W_i(1)\right)\right]^n.$$

Theorem 7.1. *For $N > 1$, $n > 1$ the inequality*

$$A(n, N) \ge e^\mu \exp\left(\sigma B_N \sqrt{n}\left(1 - \frac{\sigma B_N}{2\sqrt{n}}\right)\right) \tag{7.6}$$

is true with $B_N = M \max(W_i(1),\ 1 \le i \le N)$.

Proof. As $B_N > 0$ for $N > 1$ and

$$e^x \ge 1 + x, \quad -\infty < x < \infty$$

then

$$A(n, N) \ge e^\mu \left[M\left(1 + \frac{\sigma}{\sqrt{n}} \max_{1 \le i \le N} W_i(1)\right)\right]^n = e^\mu \left[1 + \frac{\sigma}{\sqrt{n}} B_N\right]^n. \tag{7.7}$$

Calculating the logarithms of the formula (7.7) left and right sides, obtain that for $\sigma B_N / \sqrt{n} < 1$

$$\ln A(n, N) \ge \mu + n \ln\left(1 + \frac{\sigma B_N}{\sqrt{n}}\right) \ge$$

$$\ge \mu + n\left(\frac{\sigma B_N}{\sqrt{n}} - \frac{\sigma^2 B_N^2}{2n}\right) = \mu + \sqrt{n}\sigma B_N\left(1 - \frac{\sigma B_N}{2\sqrt{n}}\right).$$

If for some n_0, $n_0 > 1$, $\sigma B_N / 2\sqrt{n_0} < 1$ then from (7.6) obtain the inequality

$$A(n, N) > (C_N)^{\sqrt{n}}, \ n > n_0, \ C_N > 1,$$

characterizing the cooperative effect in the linear stochastic growth model (7.1), (7.2).

7.2.   Basis and development of algorithm of mutual debts discharge solution

The problems of the debts discharge in very important for Russian economics. A direct way of its solution is connected with manifold technical difficulties: a complexity of an operative information collection, an information falsity, an insufficient velocity of a reaction on a fast varying situation and etc. Usual methods of the discharge are too far from optimal.

Nevertheless it is possible to improve discharge procedure so that it may become simple and sufficiently convenient for an adjustment of relations in a corporation and for a restoration of normal domesticities between Russian regions and etc. The last question is especially important because the group of main Russian economists suggested to get over economical crisis by the discharge of the arrears.

Consider the following algorithm of the debts discharge. Suppose that there are n corporators or persons $L_1, \ldots, L_n$ which should like to discharge their mutual debts without additional finances. This problem was considered in [40, 41, 77] where different approaches are suggested. Especially mark the paper [77], in which the algorithm similar to the solution of the general transportation problem is suggested. In a comparison with [77] in [3] there is a consideration of the debts discharge problem solutions as permissible solutions of some partial transportation problem. So there is a possibility to use mathematical methods which are well know for economists.

Suppose that $A = \|a_{ij}\|_{i,j=\overline{1,n}}$ is debts matrix, where a_{ij} is the debt of the $i-$th person to the $j-$th person. Without the restriction of the generality take $a_{ij} \geq 0$, $a_{ii} = 0$. Denote $z_i = a_{i1} + \ldots + a_{in}$, $y_j = a_{1j} + \ldots + a_{nj}$. Call the person L_i the creditor if $v_i = y_i - z_i \geq 0$ and the debtor in the opposite case. Suppose that first m $(m \leq n)$ persons are creditors and other ones - debtors. It is obvious that

$$v_1 + \cdots + v_m = -v_{m+1} - \cdots - v_n.$$

Our aim is to construct the outgoings matrix $B = \|b_{ij}\|$, $b_{ij} \leq a_{ij}$, $i, j = 1, \ldots, n$, where the element b_{ij} is the recommended outgoing of the $i-$th person to the $j-$th person. The matrix B is to possess two following properties :

1) The sum of the elements in the $i-$th line is to be equal to the sum of the elements in the $i-$th column that is each person pays the sum equal to the sum, which this

person receives from another persons;

2) The sum of the elements in the matrix $C = \|c_{ij}\|$, $c_{ij} \geq 0$, $i, j = 1, \ldots, n$, of the remained debts is to be minimal.

It is more convenient to consider the matrix C. Divide all debts into four groups: the debts of the creditors to the creditors ($K \to K$), the debts of the creditors to the debtors ($K \to D$), the debts of the debtors to the creditors ($D \to K$) and the debts of the debtors to the debtors ($D \to D$). All elements in the matrix C, corresponding all kinds of the debts besides of ($D \to K$), put equal zero that is $c_{ij} = 0$ for $i = m + 1, \ldots, n$, $j = m + 1, \ldots, n$. Nonzero elements may be only the elements, which correspond to the debts $D \to K$ that is c_{ij} for $i = m + 1, \ldots, n$, $j = 1, \ldots, m$. It is necessary to choose these debts so that the sum of the i−th line elements equals to $-v_i$ and the sum of the j−th column equals to v_j. This procedure gives the permissible solution of the transportation problem in its the simplest variant (when all tariffs are equal). Call the matrices constructed in this way by the matrices of the type (c). Prove that the matrix B obtained from the equality $B = A - C$ with by the solution of this transportation problem minimizes as the total sum of the debts so the sum of the debts of each person and the sum of the debts to each person.

It is known that the transportation problem solution exists [27] and contains not more than ($n-1$) nonzero elements. Define the coefficient of the limit effectiveness k as the ratio between the number of the cancelled debts to maximal possible number of the initial debts. Simple calculations showed that: $1 - 1/n \leq k \leq 1$. So the number k/n approximately equals to one.

This method is simple and convenient and clear for practicals. So it is worthy to analyze an influence of different factors on the discharge procedure. These factors are an application of the discharge to avoid the taxes. This problem may be solved by the addition of taxes to the debts and so may be reduced to the initial one.

In a practice there are situations when the discharge is in frames of some isolated coalitions or in frames of coalitions which exchange with each other via agents. This problem influences on the discharge effectiveness and is to take into account at initial stage of the discharge. So it is worthy to consider an application of the barter in the debts discharge. Another problem is to take into account an influence of errors in an initial information. Once more restriction on the debts discharge is an information confidentiality. In this subsection a way of data processing which makes more difficult the information circulation is considered.

The first step of the suggested algorithm is a replacement of the initial problem by some transportation problem. It is equivalent to the calculation of the persons balances and practically does not demand computer time. Second and main step is to use the potential method of the transportation problem solution. Attempts to refuse from this problem and to use the linear programming problem do not give good results because of large computations volumes.

Optimal theorem for remained debts matrix

Say that $X = \| x_{ij} \|_{i,j=1}^n$ is the debts matrix if

$$x_{ii} = 0, \quad x_{ij} \geq 0, \quad 1 \leq j \leq n.$$

Denote the total debt of the i-th subject by $S_i(X)$, the total debt to the i-th subject by $S^i(X)$ and the total debt by $S(X)$ via the equalities

$$S_i(X) = \sum_{j=1}^n x_{ij}, \quad S^i(X) = \sum_{j=1}^n x_{ji}, \quad 1 \leq i \leq n,$$

$$S(x) = \sum_{i=1}^n S_i(X) = \sum_{i=1}^n S^i(X).$$

Define the set $\mathcal{X}(A)$ of the permissible debts matrices X by the condition of the balances conservation for all subjects

$$S^i(X) - S_i(X) = S^i(A) - S_i(A) = v_i, \quad 1 \leq i \leq n. \tag{7.1}$$

Put the problem (a) to minimize at the set $\mathcal{X}(A)$ of all permissible dents matrices X the functions $S_i(X)$, $S^i(X)$, $1 \leq i \leq n$, and the sum $S(X)$.

Theorem 7.1. *The problem (a) solutions are only the matrices (c). That is these matrices are to construct as the permissible solutions of the corresponding transportation problem.*

Proof. It is obvious that for any matrix $X \in \mathcal{X}(A)$

$$S^i(X) \geq \max(0, v_i), \quad S_i(X) \geq \max(0, -v_i), \quad 1 \leq i \leq m. \tag{7.2}$$

Each matrix X of the type (c) satisfies the inclusion $X \in \mathcal{X}(A)$ and transforms the inequalities (7.2) into the equalities

$$S_i(X) = \max(0, v_i), \quad S^i(X) = \max(0, -v_i),$$

$$S(X) = \sum_{i=1}^{n} \max(0, v_i). \tag{7.3}$$

So the matrix X of the type (c) is the solution of the problem (a) and the equalities (7.3) are true. Conversely if the matrix X is the problem (a) solution then the equalities (7.3) are true. So the matrix X has the form (c).

Addition of debts matrices

Suppose that the initial debts matrices A', A'' correspond to the balances vectors $(v_1', \ldots, v_n')$, $(v_1'', \ldots, v_n'')$.

Theorem 7.2. *Following inequalities are true:*

$$\min_{X \in \mathcal{X}(A'+A'')} S_i(X) \leq \min_{X \in \mathcal{X}(A')} S_i(X) + \min_{X \in \mathcal{X}(A'')} S_i(X), \; 1 \leq i \leq n, \tag{7.4}$$

$$\min_{X \in \mathcal{X}(A'+A'')} S^i(X) \leq \min_{X \in \mathcal{X}(A')} S^i(X) + \min_{X \in \mathcal{X}(A'')} S^i(X), \; 1 \leq i \leq n, \tag{7.5}$$

$$\min_{X \in \mathcal{X}(A'+A'')} S(X) \leq \min_{X \in \mathcal{X}(A')} S(X) + \min_{X \in \mathcal{X}(A'')} S(X). \tag{7.6}$$

Proof. It is obvious that the sum of the matrices $A' + A''$ corresponds to the sum of the balances vectors:

$$(v_1' + v_1'', \ldots, v_n' + v_n'').$$

The solution of the problem (a) for the debts matrices A', A'', $A' + A''$ accordingly to the formula (7.3) leads to

$$\min_{X \in \mathcal{X}(A')} S_i(X) = \max(0, v_i'), \quad \min_{X \in \mathcal{X}(A'')} S_i(X) = \max(0, v_i''),$$

$$\min_{X \in \mathcal{X}(A'+A'')} S_i(X) = \max(0, v_i' + v_i''), \quad 1 \leq i \leq n.$$

Compare three last formulas and obtain the inequalities (7.4). The inequalities (7.5) are proved analogously. The inequality (7.6) is obtained by the addition of the inequalities (7.5) or (7.6) with $i = 1, \ldots, n$.

So the theorem 7.2 gives that the debt calculated by the sum matrix is smaller than the sum of the corresponding quantities calculated by the summands matrices. This remark is important in a division of the debt into different summands like main debt, percent debt and etc with the next discharge of each of them separately.

Estimate of error in discharge problem

Important corollary in the theorem 7.2 is an estimate of an error of the discharge procedure in a case of inaccurate initial data discussed at the introduction to this subsection in details. Suppose that the discharge was made accordingly to the matrix A' which equals to the matrix A but disturbed by measurements errors. Then the remained sum of the debts may be estimated as follows.

Theorem 7.3. *Following inequalities are true:*

$$\min_{X \in \mathcal{X}(A')} S(X) \leq \min_{X \in \mathcal{X}(A)} S(X) + \sum_{i=1}^{n} \max(0,\ v_i' - v_i), \tag{7.7}$$

$$\max\left(0,\ \min_{X \in \mathcal{X}(A)} S(X) - \sum_{i=1}^{n} \max(0,\ v_i' - v_i)\right) \leq \min_{X \in \mathcal{X}(A')} S(X). \tag{7.8}$$

Proof. Using the equalities (7.3), obtain

$$\min_{X \in \mathcal{X}(A)} S(X) = \sum_{i=1}^{n} \max(0,\ v_i),$$

$$\min_{X \in \mathcal{X}(A')} S(X) = \sum_{i=1}^{n} \max(0,\ v_i') \leq \sum_{i=1}^{n} \left(\max(0,\ v_i) + \max(0,\ v_i' - v_i)\right).$$

So the inequality (7.7) is true. The inequality (7.8) is obtained from the inequality (7.7) with the replacement of the matrix A by the matrix A' and vice-versa and with the obvious formula

$$\sum_{i=1}^{n} v_i = \sum_{i=1}^{n} v_i' = 0.$$

Remark 1. *Emphasize that the inequality (7.7) transforms into the equality if*

$$sign(v_i' - v_i) = sign(v_i), \quad 1 \leq i \leq n.$$

Permissible disturbances of debts matrices

The theorem 7.1 affirms that the optimal matrix of the remained debts is defined only by the balances of the separate subjects v_i, $1 \le i \le n$, calculated by the matrix A of the initial debts. So the matrix A contains the information, which is excess for the optimal debts matrix construction. And a disclosure of this information very often is undesirable. The problem is to construct the additive corruptions $Y = \|y_{ij}\|_{i,j=1}^{n}$ of the initial debts matrix A so that the balances of the separate subjects in the matrix $A' = A + Y$ satisfy the conditions:

$$v_i' = v_i, \quad i = 1, \ldots, n. \tag{7.9}$$

Denote by $\mathcal{Y}$ the set of all possible corruptions Y, satisfying the equalities (7.9). Say that these corruptions are permissible. Choose the elements y_{nn}, y_{ij}, $i = \overline{1, n}$, $j = 1, \ldots, n-1$, of the matrix Y arbitrarily. Define the remained elements y_{jn} by the equalities

$$y_{jn} = \sum_{k=1}^{n} y_{kj} - \sum_{k=1}^{n-1} y_{jk} = S^j(Y) - \sum_{k=1}^{n-1} y_{jk}, \quad j = 1, \ldots, n-1. \tag{7.10}$$

Theorem 7.4. *The set of all possible corruptions $\mathcal{Y}$ consists of the matrices (7.10) and only of them.*

Proof. Rewrite the equalities (7.9) in the more convenient form:

$$S_j(Y) = S^j(Y), \quad j = 1, \ldots, n. \tag{7.11}$$

Suppose that the elements of the matrix Y satisfy the equalities (7.10) and consequently obtain the equalities (7.11) for $j = 1, \ldots, n-1$. As

$$S^n(Y) - S_n(y) = \sum_{j=1}^{n-1} \left(S^j(Y) - \sum_{k=1}^{n-1} y_{jk} \right) + y_{nn} - \sum_{j=1}^{n} y_{nj} =$$

$$= \sum_{j=1}^{n-1} S^j(Y) - \sum_{j=1}^{n-1}\sum_{k=1}^{n-1} y_{jk} - \sum_{j=1}^{n-1} y_{nj} = \sum_{j=1}^{n-1} S^j(Y) - \sum_{j-1}^{n-1} S^j(Y) = 0.$$

So the equality (7.11) is true for $j = n$ and $Y \in \mathcal{Y}$. Conversely if $Y \in \mathcal{Y}$ and so the equalities (7.11) are true then for fixed y_{nn}, y_{ij}, $i = 1, \ldots, n$, $j = 1, \ldots, n-1$, the equalities (7.10) are true.

The theorem 7.4 allows to establish that $\mathcal{Y}$ is the linear space with the dimension $n^2 - n + 1$ and gives the possibility to calculate the elements of these space by the formulas(7.10). The formulas (7.10) may be obtained as a corollary of the well known in the transportation problem [27] fact of the linear dependence of the equations (7.11). More detailed this problem was considered in [78].

Discharge in isolated coalitions

Suppose that the initial set of subjects $I = \{1, \ldots, n\}$ is divided into m the coalitions $I_1, \ldots, I_m$ with the strengths $n_1, \ldots, n_m$ correspondingly

$$\sum_{i=1}^{m} n_i = n, \ \ I_k = \{r_{k-1} + 1, \ldots, r_k\}, \ \ \ k = \overline{1, m},$$

$$r_0 = 0, \ r_k = r_{k-1} + n_k, \ \ k = \overline{1, m}.$$

Debts discharge in the isolated coalition I_k is the discharge for the submatrix

$$A_k = \|a_{ij}\|, \ \ \ r_{k-1} + 1 \leq i, j \leq r_k,$$

of the matrix A. Replace in the matrix A the square submatrices A_k by the submatrices, obtained after the discharge in the separate coalitions I_k, $k = 1, \ldots, m$. Denote such matrix by A_*^c. Define the matrix A_* as the matrix of the remained debts, which is obtained after the complete discharge of the initial matrix A. The theorem 7.1 leads that the matrices A_*^c, A_* are defined ambiguity but this ambiguity does not influence on next constructions.

Define the matrix A^e, obtained from A by zeroing all elements in the submatrices A_k, $k = 1, \ldots, m$. It is obvious that A^e may be considered as the matrix of an external debts of the coalitions. Suppose that A_*^e is the matrix of the debts, remained after the discharge of the matrix A^e of the external debts. Denote

$$d(A^e) = S(A^e) - S(A_*^e) \geq 0,$$

the quantity $d(A^e)$ characterizes the effect of the discharge by the external debts matrix.

Theorem 7.5. *The inequality*

$$S(A_*^c) \geq S(A_*) + d(A^e) \tag{7.12}$$

is true.

Proof. The theorem 7.5 proof is based on the following statement.

Lemma 7.1. *Suppose that* $\alpha, \beta, \gamma, \delta$ *are the arbitrary nonnegative numbers then*

$$(\alpha - \beta)^+ + \gamma \geq ((\alpha + \gamma) - (\beta + \delta))^+ + \gamma - (\gamma - \delta)^+ \tag{7.13}$$

with $u^+ = \max(0, u)$.

Proof. For any arbitrary $u, v, -\infty < u, v < \infty$ the following inequalities are true

$$u^+ + v^+ \geq u^+ + v \geq (u + v)^+. \tag{7.14}$$

Choose now $u = \alpha - \beta, \quad v = \gamma - \delta$ then accordingly to the formula (7.14) obtain

$$(\alpha - \beta)^+ \geq ((\alpha - \beta) + (\gamma - \delta))^+ - (\gamma - \delta)^+. \tag{7.15}$$

Add γ to the both sides of the inequality in the formula (7.15) and obtain the formula (7.13).

Using the lemma 7.1 and the formulas (7.3) obtain

$$S(A_*^c) = \sum_{k=1}^{m} \sum_{i \in I_k} [(S_i(A_k) - S^i(A_k))^+ + S_i(A^e)] \geq$$

$$\geq \sum_{k=1}^{m} \sum_{i \in I_k} [((S_i(A_k) + S_i(A^e)) - (S^i(A_k) + S^i(A^e)))^+ +$$

$$+ S_i(A^e) - (S_i(A^e) - S^i(A^e))^+] =$$

$$= \sum_{k=1}^{m} \sum_{i \in I_k} [(S_i(A_*) - S^i(A_*))^+ + S_i(A_*) - (S_i(A^e) - S^i(A^e))^+] =$$

$$= \sum_{i=1}^{n} [(S_i(A_*) - S^i(A_*))^+ + S_i(A^e) - (S_i(A^e) - S^i(A^e))^+] =$$

$$= S(A_*) + S(A^e) - S(A_*^c) = S(A_*) + d(A^e).$$

Consequently the formula (7.12) is true.

So accordingly to the formula (7.12) the discharge in the isolated coalitions (that is by the matrix A_*^c) leads to an increasing of the remained debts sum in a comparison with the discharge made without the coalitions (that is by the matrix A_*) at least with the quantity $d(A_*^e)$. This quantity characterizes the effect of the discharge by the matrix A^e.

Discharge over coalitions agents

In a practice there is a lot of discussions and conflicts connected with agents role in the discharge. Naturally a presence of the agents is estimated negatively. But an attempt to refuse from the agents does not give positive results too. So there is a problem to define necessary involvement of the agents in the discharge. This problem solution is convenient to make in frames of the suggested discharge procedure in the presence of some coalitions.

Consider some possible way of the discharge via the coalitions agents. This way is based on the calculation of the balance V_k of the k-th coalition I_k,

$$V_k = \sum_{i \in I_k} v_i,$$

and an incorporation into I_k the fictitious subject i^k, which has the internal balance $-V_k$ and the external balance V_k, $k = 1, \ldots, m$. The discharge is made separately in the incorporated coalitions

$$\overline{I}_k = \{r_{k-1} + 1, \ldots, r_k, i^k\}, \quad k = 1, \ldots, m,$$

and in the group of the agents $\{i^1, \ldots, i^m\}$. In each of these sets the discharge actually is made not by the matrix of the initial debts but only by the balances vector. From the formula (7.3) obtain that the sum of the components of the balances vector constructed by the debts matrix equals zero.

Calculate the sum of the balances of the subjects from the set $\overline{I}_k$:

$$\sum_{i \in I_k} v_i + (-V_k) = V_k - V_k = 0, \quad k = 1, \ldots, m. \tag{7.16}$$

Calculate the sum of the balances of the subjects from the group of the agents:

$$\sum_{k=1}^{m} V_k = \sum_{k=1}^{m} \sum_{i \in I_k} v_i = \sum_{i=1}^{n} v_i = 0. \tag{7.17}$$

So if we know only the balances of the subjects from the sets $\{I_1, \ldots, I_m\}, \{i^1, \ldots, i^m\}$ then it is possible, using the formulas (7.16), (7.17), to construct the matrices of the mutual debts for each set and to define the matrix of the remained debts in the set $\overline{I} = \overline{I}_1 \cup \ldots \cup \overline{I}_m$. It is known that the construction of the remained debts matrix by the balances vector is multi valued procedure but the sums of the separate subjects remained debts are defined by these balances uniquely (7.3). Analogous statement is true for the sums of the separate subjects debts.

Theorem 7.6. *Minimal sum of the remained debts* $S(A_*)$ *in the procedure of the discharge without the agents differs from the sum of the remained debts* $S(\overline{I})$ *in the procedure with the agents by the quantity* $\sum\limits_{k=1}^{m} |V_k|$:

$$S((\overline{I}) - S(A_*) = \sum_{k=1}^{m} |V_k|.$$

Proof. As each initial subject i, $1 \leq i \leq n$, comes in unique coalition so the sum S_i of the i−th subject remained debts, calculated by the suggested enhanced matrix, satisfy the obvious equalities

$$S_i = v_i^+, \quad S^i = (-v_i)^+, \quad i = 1, \ldots, n. \tag{7.18}$$

Each agent i^k of the coalition I_k comes in the discharge procedure into two subject sets: $\overline{I}_k$ and $\{i^1, \ldots, i^m\}$. The sum of the remained debts of the subject i^k in the coalition $\overline{I}_k$ equals to $(-V_k)^+$ and the sum of the remained debts of the subject i^k in the set $\{i^1, \ldots, i^m\}$ equals to V_k^+. So the total sum of the subject i^k debts equals to

$$(-V_k)^+ + V_k^{'} = |V_k|. \tag{7.19}$$

The formulas (7.18), (7.19) lead that the sum of all debts $S(\overline{I})$ in the set $\overline{I}$ of all subjects satisfies the equality

$$S(\overline{I}) = \sum_{i=1}^{n} v_i^+ + \sum_{k=1}^{m} |V_k| = S(A_*) + \sum_{k=1}^{m} |V_k|. \tag{7.20}$$

Suggested algorithm of the debts discharge via the coalitions agents obviously permits a repetition but in the coalition of the agents. If all the discharge procedures in the coalitions (of the initial subjects or agents) are made using the North-West angle method [27] then the number of the remained nonzero debts is not larger than the number of all participants minus one. Then the total number of nonzero remained debts in the procedure of the discharge via the agents is not larger than $N_1 \leq n + m - 1$. If the group of m agents of the initial coalitions is divided into

m_1 subcoalitions and the discharge into them is made using the previous algorithm then the total number of the nonzero remained debts N_2 satisfies the inequality $N_2 \le n + m + m_1 - 1$ and so on. More detailed the problems of last two points are considered in [79].

Discharge over barter

In a practice the debts discharge may be realized as by money so by barter. As practicals have an interest to this form of the discharge so it is worthy to consider the barter in frames of some mathematical model, in which money is replaced by an article. Investigate how does this suggestion influence on an effectiveness of the discharge.

Suppose that the discharge is made accordingly to the initial debts matrix A and all subjects are divided into the debtors and the creditors accordingly to their balances $v_1, v_2, \ldots, v_n$. Without the restriction of the generality suppose that the debtors are the 1-st, the 2-nd,..., the k-th subjects and all other subjects are the creditors. Denote $u_j = -v_{k+j}, \quad 1 \le j \le m$.

At first consider a possibility of the debts discharge by the articles of one type. Suppose that debtors have $d_i, \ 1 \le i \le k$, units of this article. And the creditors may take on an account of the debts not larger than $D_j, \ 1 \le j \le m$ units of the article. the unit article prize at a market is c. In these restrictions it is necessary to calculate the matrix $Z = \|z_{ij}\|_{i=1, j=1}^{k, \ m}$ of the articles volumes, which are to be sent from the debtors to the creditors, so that remained debts sum is minimal. This problem may be considered mathematically as a search of the matrix Z so that

$$
\begin{aligned}
&cs(Z) \rightarrow \max, \\
&s_i(Z) \le d_i, \quad cs_i(Z) \le v_i, \quad 1 \le i \le k, \\
&s^j(Z) \le D_j, \quad cs^j(Z) \le u_j, \quad 1 \le j \le m,
\end{aligned}
\tag{7.21}
$$

where

$$
s_i(Z) = \sum_{j=1}^{m} z_{ij}, \ 1 \le i \le k, \quad s^j(Z) = \sum_{i=1}^{k} z_{ij}, \ 1 \le j \le m,
$$

$$
s = s(Z) = \sum_{i=1}^{k} s_i(Z).
$$

The problem (7.21) may be modified if we suggest that the price c is the function of the total volume of the sent article s :

$$c = f(s). \tag{7.22}$$

So the formula (7.22) permits not only a market but a wholesale price which decreases when s increases accordingly to some law. Here $f(s)$, $s > 0$, is some function, which is continuous and monotonically decreasing by s.

Denote that as the problem (7.21) so its modification (7.22) take into account only the sums of separate lines and columns in the matrix Z. That is as the initial problem (a) of the debts discharge so the problem (7.21), (7.22) of the debts discharge with a help of the article contain only sum but not paired disparity characteristics. This remark allows to find sufficiently simple cooperative solutions of this problem.

At first consider the problem (7.21) in which c is fixed. From the conditions of this problem it is easy to obtain that

$$s \leq \min \left(\sum_{i=1}^{k} \min \left(d_i, \frac{v_i}{c} \right), \ \sum_{j=1}^{m} \min \left(D_j, \frac{u_j}{c} \right) \right). \tag{7.23}$$

Construct the matrix Z by the North-West method [27] with the restrictions

$$s_i(Z) \leq \min \left(d_i, \frac{v_i}{c} \right), \quad 1 \leq i \leq k,$$

$$s^j(Z) \leq \min \left(D_j, \frac{u_j}{c} \right), \quad 1 \leq j \leq m.$$

It is easy to show that the matrix $Z = Z(c)$, constructed by this method, converts the inequality (7.23) into the equality ($c > 0$):

$$s = \min \left(\sum_{i=1}^{k} \min \left(d_i, \frac{v_i}{c} \right), \ \sum_{j=1}^{m} \min \left(D_j, \frac{u_j}{c} \right) \right) = g(c). \tag{7.24}$$

The matrix $Z(c)$ of the article supplies contains not more than $k + m - 1 = n - 1$ nonzero elements. The equalities (7.24) convert the problem (7.21), (7.22) into the solution of following equations system:

$$c = f(s), \quad s = g(c). \tag{7.25}$$

If the vector (s, c) is the system (7.25) solution then $Z = Z(c)$ is the solution of the problem (7.21), (7.22) with $s(Z(c)) = s$. An apparent appearance of the function $g(c)$, introduced in the formula (7.24), significantly enables the system (7.25) analysis. More detailed this problem was considered in [80].

Discharge with nonnegative outgoings

Another restriction on the discharge procedure is the demand (b) of the matrix B of the outgoings nonnegativity. The introduction of the condition (b) into the problem (a) gives the problem (ab), which can not be considered as the transportation problem and may be considered only in frames of the complete linear programming problem. In this problem as the calculations time so the sum of the matrix C elements (the matrix of remained debts) increase. It is shown that for the triangle matrix A of the initial debts the problem (ab) has zero matrix of the outgoings.

Suppose that P, Q are square matrices with r columns, $r > 0$. Say that $P \geq Q$ if all elements of the matrix $P - Q$ are nonnegative. Put $P > Q$ if $P \geq Q$ and there is at least one nonzero element in the matrix $P - Q$. It is obvious that the square matrices A, B, C with the size $n \times n$ satisfy the conditions:

$$A \geq \mathbf{0}, \quad C \geq \mathbf{0}, \quad A = B + C \Longrightarrow A \geq B,$$
$$S_i(B) = S^i(B), \quad 1 \leq i \leq n, \tag{7.26}$$

in which $\mathbf{0}$ is zero matrix.

Theorem 7.7. *Suppose that the matrix A of the initial debts is triangle and the condition (7.26) is true then*

$$B \geq \mathbf{0} \Longrightarrow B = \mathbf{0}. \tag{7.27}$$

Proof. Without the restriction of the generality suppose that all over diagonal and diagonal elements of the matrix A are zeros and so

$$a_{1\,n} = a_{2\,n} = \ldots = a_{n\,n} = 0.$$

Accordingly to (7.26) $A \geq B$ and consequently for $B \geq \mathbf{0}$

$$b_{1\,n} = b_{2\,n} = \ldots = b_{n-1\,n} = b_{n\,n} = b_{n\,n-1} = \ldots = b_{n\,2} = b_{n\,1} = 0 \tag{7.28}$$

as $S_1(B) = S^1(B)$. So the last line and the last column in the matrix B consist of zeros.

Construct now from the matrices A, B the matrices A_1, B_1 by deleting in A, B the last lines and columns. As $A \geq B \geq \mathbf{0}$ so $A_1 \geq B_1 \geq \mathbf{0}$. From (7.26), (7.28) obtain

$$S_i(B_1) = S^i(B_1), \quad 1 \leq i \leq n - 1.$$

So $A_1 \geq B_1 \geq \mathbf{0}$ and the matrix B_1 may be considered as the matrix of the outgoings for the matrix A_1. Repeating these considerations, obtain

$$b_{1\,n-1} = b_{2\,n-1} = \ldots = b_{n-2\,n-1} = b_{n-1\,n-1} =$$
$$= b_{n-1\,n-2} = \ldots = b_{n-1\,2} = b_{n-1\,1} = 0.$$

Repeating this consideration again $n - 3$ times, obtain the formula (7.27).

Remark 2. *In the theorem 7.7 it is possible to replace the triangle condition for the matrix A by more general one: there is the permutation of the subjects numbers $1, \ldots, n$ so that obtained matrix becomes triangle. The set of the matrices $\{A\}$, satisfying this condition, denote by $\mathcal{A}^*$.*

Theorem 7.8. *Suppose that A is the initial debts matrix and the conditions (7.26), (7.27) are true then $A \in \mathcal{A}^*$.*

Proof. Suppose that the conditions (7.26), (7.27) are true but the matrix A does not contain zero lines. Take in each line of the matrix A nonzero element. Define by these elements the matrix A', $A \geq A'$, in which only taken elements are nonzero and all other elements are zeros. Denote by I' the set of numbers of zero columns in the matrix A'. Consider the case when I' is nonempty set.

Delete from the matrix A' the lines and the columns with the numbers belonging to I'. Remained lines and columns conserve previous numeration. Repeat this procedure again and again to remove in obtained matrix $\overline{A}$ zero columns. It is obvious that the number of these repetitions is finite. Then the final matrix $\overline{A}$ has in each column and in each line single positive element. Denote by $\bar{a}$ the minimum of positive elements in the matrix $\overline{A}$. Take initial matrix A and conserve all nonzero elements of the matrix $\overline{A}$ (accordingly to their initial indexes), all other elements replace by zeros. Denote obtained matrix by $\overline{\overline{A}}$.

Construct now the matrix B, putting in it at the places of nonzero elements of the matrix $\overline{\overline{A}}$ the number $\overline{a}/2$ and zeros at all other places. It is clear that $A \geq \overline{\overline{A}} \geq B > \mathbf{0}$ and $S_i(B) = S^i(B)$, $1 \leq i \leq n$. So B is the matrix of the outgoings, $A \geq B > \mathbf{0}$. The contradiction with the theorem conditions leads that the matrix A has zero line. Renumber the subjects $1, 2, \ldots, n$ so that the n–th line of obtained matrix is zero. Analogously with the theorem 7.7 proof proceed from A, B to the matrices A_1, B_1 and repeat this reasoning. As a result find in the matrix A_1 zero line. Without the restriction of the generality suppose that it is last $(n-1)$-th line. Repeating $n-3$ times this reasoning, obtain that after some renumbering of the indexes $1, 2, \ldots, n$ the matrix A becomes triangle: $A \in \mathcal{A}^*$.

7.3.   Qualitative investigation of Estes self-education model

Mathematical models of an adaptation in a presence of some possible choice variants are widely used in the ecology, the ethology the psychology and the economics [17, 68, 4, 5, 22]. A fortune of these models is defined by the complexity of their mathematical investigation. For an example the adaptation models considered in [4, 5] are based on the classical "urns" scheme of the probability theory and give sufficiently apparent and easily interpreted results based on properties of ergodic Markov chains. At the same moment the adaptation models considered in the papers [17, 68, 22] lead to the Markov chains with limit distributions, depending on their initial states. This dependence is described by nonstandard functional equation. Last circumstance created some difficulties in an interpretation of the model results.

This subsection is devoted to an investigation of this equation solution with a purpose to obtain apparent interpretation of the solutions of analyzed adaptation model.

Consider mathematical model of predator adaptation to two kinds of possible food with approximately equal quality [68]. Suppose that after the usage of one food kind the interest to this food kind increases. Mathematical model of such adaptation (self- education) is suggested by Estes [22] and is based on the following suggestions. Suppose that p_n is the probability that the 1-st food kind is chosen, $q_n = 1 - p_n$ is the probability that the 2-nd food kind is chosen at the n–th step,

$n = 1, 2, \ldots$ Denote by C_i the coefficient of the self-education to the i−th kind of the food, $0 < C_i < 1$ (usually for animals $C_i \leq 0.1$), $i = 1, 2$. If on the n−th step the 1-st food kind is chosen then

$$p_{n+1} = p_n + C_1(1 - p_n), \quad q_{n+1} = 1 - p_{n+1} = (1 - C_1)q_n, \tag{7.29}$$

if on the n−th step the 2-nd food kind is chosen then

$$q_{n+1} = q_n + C_2(1 - q_n),$$

$$p_{n+1} = 1 - q_{n+1} = (1 - C_2)p_n, \ n = 1, 2, \ldots. \tag{7.30}$$

Suppose that all choices are independent then the sequence p_n, $n = 1, 2, \ldots$, is the Markov chain. The formulas (7.29), (7.30) are based empirically in [13].

The aim of this subsection is the investigation of the function

$$F(x, C_1, C_2) = P(\lim_{n \to \infty} p_n = 0/p_1 = x), \ 0 \leq x \leq 1, \tag{7.31}$$

for small coefficients of the education abilities C_1, C_2. Main result of this investigation is the following statement. For small C_1, C_2 dependently on the ratio between C_1, C_2 the function $F(x, C_1, C_2)$ may varies in the interval $(\varepsilon, \ 1 - \varepsilon)$, where $0 < \varepsilon < 1/2$, or uniformly tends to 0 or uniformly tends to 1 or coincides with the function $1 - x$. This behavior of the function gives clear interpretation of the model (7.29), (7.30). Obtained results are comparable with analogous results obtained from the urns model [4, 5], which found applications in mathematical economics.

Auxiliary results

Denote $f(x, h, k) = F(x, C_1, C_2)$ where $h = 1 - C_1$, $k = 1 - C_2$. Fix h, k, $0 < h < k < 1$, and designate $f(x, h, k) = g(x)$. Using complete the probability formula and (7.29), (7.30), (7.31), it is easy to obtain for the function $g(x)$ [68] the equation

$$g(x) = xg(1 - h(1 - x)) + (1 - x)g(kx) \tag{7.32}$$

with the boundary conditions

$$g(0) = 1, \quad g(1) = 0. \tag{7.33}$$

Suppose that $C[0, 1]$ is the space of the continuous functions on the set $[0, 1]$ with the norm $\|c(x)\| = \sup_{0 \leq x \leq 1} |c(x)|$. In [13], [17, p. 11–13] it is proved that for arbitrary h, k, $0 < h < k < 1$, the problem (7.32), (7.33) has in the space $C[0, 1]$ the unique solution and there exists the function $\psi(x) \in C[0, 1]$ so that

$$g(x) = (1 - x)\alpha(x, k) + hx(1 - x)\psi(x),$$
(7.34)

$$\alpha(x, k) = \prod_{n=1}^{\infty} (1 - k^n x),$$
(7.35)

$$\|\psi(x)\| \leq \frac{k}{(k - h)(1 - k)} \|\alpha(1 - h(1 - x), k)\|.$$
(7.36)

Lemma 7.2. *The following inequality is true:*

$$\alpha(x, k) \leq \exp\left(-\frac{kx}{1 - k}\right), \quad 0 \leq x \leq 1, \quad 0 < k < 1.$$
(7.37)

Proof. It is obvious that for $u \geq 0$ the inequality

$$1 - u \leq e^{-u}$$
(7.38)

is true. Combining the formulas (7.35), (7.38), obtain

$$\alpha(x, k) \leq \prod_{n=1}^{\infty} \exp(-k^n x) \leq \exp\left(-\sum_{n=1}^{\infty} k^n x\right) = \exp\left(-\frac{kx}{1 - k}\right),$$
$$0 \leq x \leq 1, \quad 0 < k < 1.$$

The formula (7.37) is proved.

Lemma 7.3. *The inequality*

$$\|\alpha(1 - h(1 - x), k)\| \leq \exp\left(-\frac{k(1 - h)}{1 - k}\right), \quad 0 < k, h < 1,$$
(7.39)

is true.

Proof. Accordingly to the lemma 7.2

$$|\alpha(1 - h(1 - x), k)| = \alpha(1 - h(1 - x), k) \leq$$
$$\leq \exp\left(-\frac{k(1 - h(1 - x))}{1 - k}\right) \leq \exp\left(-\frac{k(1 - h)}{1 - k}\right),$$
$$0 \leq x \leq 1, \quad 0 < k, h < 1.$$

So the inequality (7.39) takes place.

Lemma 7.4. *The inequality*

$$f(x, h, k) \le \exp\left(-\frac{kx}{1-k}\right) + \frac{1}{(k-h)(1-k)} \exp\left(-\frac{k(1-h)}{1-k}\right), \tag{7.40}$$

$$0 \le x \le 1, \quad 0 < h < k < 1,$$

is true.

Proof. Accordingly to the formula (7.34)

$$f(x, h, k) \le \alpha(x, k) + \|\psi(x)\|. \tag{7.41}$$

The formulas (7.36). (7.39) lead to

$$\|\psi(x)\| \le \frac{1}{(k-h)(1-k)} \exp\left(-\frac{k(1-h)}{1-k}\right). \tag{7.42}$$

Put the inequality (7.41) into the estimates (7.37), (7.42) and obtain the formula (7.40).

Lemma 7.5. *If $0 < \varepsilon \le x \le 1, \quad 0 < h < k < 1$ then*

$$f(x, h, k) \le \exp\left(-\frac{k\varepsilon}{1-k}\right) + \frac{1}{(k-h)(1-k)} \exp\left(-\frac{k(1-h)}{1-k}\right). \tag{7.43}$$

Proof. the inequality (7.43) is the corollary of the lemma 7.4.

Rewrite the inequality (7.43) in terms of the coefficients $C_1 = 1 - h$, $C_2 = 1 - k$ and the function $F(x, C_1, C_2) = f(x, 1 - C_1, 1 - C_2)$:

$$F(x, C_1, C_2) \le \exp\left(-\frac{\varepsilon(1 - C_2)}{C_2}\right) +$$

$$+ \frac{1}{(C_1 - C_2)C_2} \exp\left(-\frac{(1 - C_2)C_1}{C_2}\right). \tag{7.44}$$

Lemma 7.6. *Suppose that*

$$C_1 \ge -\frac{2C_2 \ln C_2}{1 - C_2}, \quad 0 < C_2 < \exp\left(-\frac{1}{2}\right) \tag{7.45}$$

then the inequality

$$F(x, C_1, C_2) \le \exp\left(-\frac{\varepsilon(1 - C_2)}{C_2}\right) + \frac{1}{-2 \ln C_2 - 1}, \quad \varepsilon < x \le 1, \tag{7.46}$$

is true.

Proof. It is obvious that the function

$$t(C_1, C_2) = \frac{1}{(C_1 - C_2)C_2} \exp\left(-\frac{(1 - C_2)C_1}{C_2}\right)$$

monotonically decreases by C_1, $0 < C_2 < C_1 < 1$. So if the condition (7.45) takes place then

$$t(C_1, C_2) = \left(\frac{-2\ln C_2}{1 - C_2} - 1\right)^{-1} \leq \frac{1}{-2\ln C_2 - 1}. \tag{7.47}$$

Combining the formulas (7.44), (7.47), obtain the inequality (7.46).

Denote

$$R(\varepsilon, C_2) = \sup\left(F(x, C_1, C_2), \ \varepsilon \leq x \leq 1 - \varepsilon, \ C_1 \geq -\frac{2C_2 \ln C_2}{1 - C_2}\right),$$

$$0 < C_2 < \exp\left(-\frac{1}{2}\right).$$

Theorem 7.9. *The function $R(\varepsilon, C_2)$, $0 < \varepsilon < 1/2$, satisfies the condition*

$$\lim_{C_2 \to 0} R(\varepsilon, C_2) = 0 \tag{7.48}$$

and the rate convergence in the formula (7.48) is characterized by the inequality

$$R(\varepsilon, C_2) \leq \exp\left(\frac{-\varepsilon(1 - C_2)}{C_2}\right) + \frac{1}{-2\ln C_2 - 1}. \tag{7.49}$$

Proof. The statement of the theorem 7.9 is the direct corollary of the lemma 7.6.

Denote

$$G(\varepsilon, C_1) = \sup\left(1 - F(x, C_2, C_1), \ \varepsilon \leq x \leq 1 - \varepsilon, \ C_2 \geq -\frac{2C_1 \ln C_1}{1 - C_1}\right),$$

$$0 < C_1 < \exp\left(-\frac{1}{2}\right).$$

Theorem 7.10. *The function $G(\varepsilon, C_1)$, $0 < \varepsilon < 1/2$, satisfies the condition*

$$\lim_{C_1 \to 0} G(\varepsilon, C_1) = 0 \tag{7.50}$$

and the rate convergence in the formula (7.50) is given by the inequality

$$G(\varepsilon, C_1) \leq \exp\left(-\frac{\varepsilon(1 - C_1)}{C_1}\right) + \frac{1}{-2\ln C_1 - 1}. \tag{7.51}$$

Proof. The theorem 7.10 is the corollary of the theorem 7.9 and the equality

$$f(x, h, k) = 1 - f(1 - x, h, k),$$

which is true [17, c. 7] for each h, k, $0 < h < k < 1$.

Main result

Analogously with the condition (7.45) introduce the condition

$$C_2 \geq -\frac{2C_1 \ln C_1}{1 - C_1}, \quad 0 < C_1 < \exp\left(-\frac{1}{2}\right). \tag{7.52}$$

Denote: by Γ_1 the set of the vectors (C_1, C_2), $0 < C_1, C_2 < 1$, satisfying the condition (7.45), by Γ_2 the set of the vectors (C_1, C_2), $0 < C_1, C_2 < 1$, satisfying the condition (7.52), by Γ_3 the set of the vectors (C_1, C_2) so that $0 < C_1 = C_2 < 1$.

Theorem 7.11. *For each x, $0 < x < 1$, the formulas*

$$\lim_{\substack{(C_1, C_2) \to (0,0) \\ (C_1, C_2) \in \Gamma_1}} F(x, C_1, C_2) = 0, \tag{7.53}$$

$$\lim_{\substack{(C_1, C_2) \to (0,0) \\ (C_1, C_2) \in \Gamma_2}} F(x, C_1, C_2) = 1, \tag{7.54}$$

$$\lim_{\substack{(C_1, C_2) \to (0,0) \\ (C_1, C_2) \in \Gamma_3}} F(x, C_1, C_2) = 1 - x \tag{7.55}$$

are true.

Proof. The formula (7.53) is the corollary of the theorem 7.9. Analogously the formula (7.54) is the corollary of the theorem 7.10.

Suppose that $0 < C_1 = C_2 < 1$. The function $F(x, C_1, C_2)$ exists, is continuous by x [18, chapter 5] and satisfies the formulas (7.32), (7.33), the solution of the problem (7.32), (7.33) is unique in the class $C[0, 1]$. As the function $(1 - x)$ is the solution of the problem (7.32), (7.33) then

$$F(x, C_1, C_2) = 1 - x, \quad 0 \leq x \leq 1, \quad 0 < C_1 = C_2 < 1. \tag{7.56}$$

The formula (7.56) leads to the equality (7.55).

Remark 3. *The formulas (7.49), (7.51) give uniform at the interval $(\varepsilon, 1 - \varepsilon)$, $0 < \varepsilon < 1/2$, estimates of the convergence rate in the limits (7.53), (7.54), correspondingly.*

The classification of the function $F(x, C_1, C_2)$ limit behavior for $(C_1, C_2) \to 0$ dependently on the belonging of the vector to one of the parameter sets(C_1, C_2) $\Gamma_1, \Gamma_2, \Gamma_3$ is in a quality correspondence with [5, table 3]. The statements of the theorems 7.9, 7.10, 7.11 correspond with the results of numerical experiments made for Estes model in the papers [17, fig. 2], [57]. The theorems 7.9, 7.10 proofs contain the rate convergence estimates. This fact significantly eases the interpretation of Estes model.

Bibliography

1. Abramov O.V. Parametrical analysis of stochastic systems with reliability demands. M, *Science*, 1992. 176 p. (In Russian).

2. Anastasi G., Lenzini L., Mingozzi E., Hettich A., Kramling A. MAC Protocols for Wideband Wireless Local Access: Evolution Toward Wireless ATM *IEEE Personal Communications*, 1998. Vol. 5, No 5. P. 53-64.

3. Anikonov D.S., Tsitsiashvili G.Sh. Reduction of debts discharge problem to transportation problem. *Lectures of RAS*, 1997. T. 352, No 6. P. 730. (In Russian).

4. Arthur W.B., Ermoliev Yu., Kaniovski Yu. Path Dependent Processes and the Emergence of Macro-Structure European *Journal of Operational Research* Vol. 30. P. 294-303.

5. Arthur W.B. Competing Technologies, Increasing Returns, and Lock in by Historical Events *The Economic Journal*, 1989. Vol. 99. P. 116-131.

6. Asmussen, S. *Ruin Probabilities* Singapore: World Scientific, 2000. 388 p.

7. Bespalov V.M., Tsitsiashvili G.Sh. On mixing in high speed gas flow. Materials of first Russian-Korean symposium of mathematical modelling. *Vladivostok: FEB RAS*, 1992. Part 2. P. 88-96. (In Russian).

8. Borovkov A.A. Asymptotical methods of queueing theory. M.: *Science*, 1980. 384 p. (in Russian).

9. Borovkov A.A. Probability theory. M.: *Science*, 1986. 432 p. (in Russian).

10. Borovkov A.A. Course of probability theory. M.: *Science*, 1972. 288 p. (in Russian).

11. Borovkov A.A. Probability processes in queueing theory. M.: *Science*, 1972. 368 p. (in Russian).

12. Buslenko N.P., Kalashnikov V.V., Kovalenko I.N. Lectures on the theory of complex systems. M.: Sov. Radio, 1973. 440 p. (in Russian).

13. Bush P., Mosteller F. Stochastic models of learning. M.: *Physmatgiz*, 1962. (in Russian).

14. Censor Ya. Methods of images reconstruction based on decomposition in finite raws *TEER*, 1983. T. 71, No 3. P. 148-160.

15. Cramer H. Sur un nouveau theoreme - limite de la theorie des probabilites *Actual. sci. et ind.* Paris, 1938. No 736.

16. Daley D. J. The moment index of minima. Probability, statistics and seismology *J. Appl. Probab.*, 2001. Vol. 38A. P. 33-36.

17. Dmitriev A.A., Shapiro A.P. On some functional equation if learning theory/ Preprint. IACP. Vladivostok: *FESC USSR AS*, 1979. 18 p. (in Russian).

18. Dub J. Probability processes. M.: *FL*, 1956. (in Russian).

19. Embrechts P., Klüppelberg C., Mikosch T. Modelling Extremal Events. Berlin: Springer, 1997.

20. Embrechts P., Kluppelberg C., *Mikoch T. Modelling Extremal Events for Insurance and Finance.* Heidelberg: Springer, 1997.

21. Embrechts P., Veraverbeke N. Estimates for the probability of ruin with special emphasis on the possibility of large claims Insurance: *Math. & Econom.*, 1982. Vol 1. P. 55–72.

22. Estes W.K. Toward a statistical theory of learning *Psychol. Rev.*, 1950. Vol. 57. P. 94-107.

23. Feller W. Generalization of a probability limit theorem of Cramer *Trans. Amer. Math. Soc.*, 1943. Vol. 54, No 2. P. 361-372.

24. Feller W. Introduction to probability theory and its applications. M.: *World*, 1984. T. 2. 738 p. (in Russian).

25. Frisman E.Ya. Variation in population dynamics: mechanisms of transition to chaos *Bull. of the Far Eastern Branch of Russian Acad. of Sci.*, 1995. No 4. P. 92-103. (in Russian).

26. Fuk D.H., Nagaev S.V. Probability inequalities for sums of independent variables *Theory Prob. Appl.*, 1971. T. 16, vol. 4. P. 660-675. (in Russian).

27. Gabasov R., Kirillova F.M. Methods of linear programming. Minsk: *BSU*, 1978. Part 2. 239 p. (in Russian).

28. Glybin L.Ya. Intradiary cyclicity of some illnesses exhibition. Vladivostok: *FESU*, 1987. 188 p. (in Russian).

29. Gnedenko B.V., Kovalenko I.N. Introduction in queueing theory. M.: *Science*, 1966. (in Russian).

30. Gnedenko B.V. Course of probability theory. M.: *Science*, 1988. 448 p. (in Russian).

31. Goldie C.M., Kluppelberg C. *Subexponential Distributions* Preprint. Johannes Guttenberg - Universitat Mainz, 1996. No 96-1. 20 p.

32. Goncharskiy A.V., Kochikov I.V., Matvienko A.N. Reconstruction processing and analysis of images in problems of calculative diagnosis. M.: *MSU*, 1993. 140 p. (in Russian).

33. Gordon K.D., Newell G.F. *Closed Queuing Systems with Exponential Servers Oper. Research*, 1967. Vol. 15, No 2, P. 254-265.

34. Ivchenko G.I., Kashtanov V.A., Kovalenko I.N. *Queueing theory*. M.: High school, 1982. 256 p. (in Russian).

35. Ivnitskiy V.A. Theory of queueing networks. M.: *Physmathedit*, 2004. 772 p. (in Russian).

36. Jackson J.R. *Networks of Waiting Lines Oper. Res.*, 1957. Vol. 5, No 4. P. 518-521.

37. Kalashnikov V.V. Quality analysis of complex systems behavior by test functions method. M.: *Science*, 1978. 248 p. (in Russian).

38. Kalashnikov V., Norberg R. Power tailed ruin probabilities in the presence of risky investments *Stochastic Process. Appl.*, 2002. Vol. 98(2). P. 211-228.

39. Kalashnikov, V.V., Konstantinides, D. Ruin under interest force and subexponential claims: a simple treatment *Insurance: Mathematics and Economics*, 2000. Vol. 27. P. 145-149.

40. Kalitkin N.N. Problem of concerns debts discharge. *Lectures of RAS*, 1995. T. 341, No 1. P. 12-14. (in Russian).

41. Kalitkin N.N., Mikhailov A.P. Ideal solution of mutual debts discharge problem *Mathematical modelling*, 1995. T. 7, No 6. (in Russian).

42. Karlin S. Foundations of random processes theory. M.: *World*, 1971. 536 p. (in Russian).

43. Kelly F.P. *Reversibility and Stochastic Networks* John Wiley & Sons, 1979.

44. Kiefer J., Wolfowitz J. On the theory of queues with many servers *rans. Amer. Math. Soc.*, 1955. Vol. 78. P. 147-161.

45. Kovalenko I.N., Kuznetsov N.Yu., Shurenkov V.M. Random processes. Kiev: *Naukova dumka*, 1983. 366 p. (in Russian).

46. Krylenko A.V. Queueing networks with some types of customers, instant serving and customers bypasses of nodes *Problems of information transmission*, 1997. T. 33, vol. 3. P. 91-101. (in Russian).

47. Kuleshov E.L. Estimates of correlation interval *Autometrics*, 1984. No 2. P. 100–102. (in Russian).

48. Kuleshov E.L. Estimate of correlation interval as functional of square type Automatization of experiment and data processing. Vladivostok: *FESU*, 1986. P. 36-42.

49. Lindley D.V. The theory of queues with a single server Proc. Camb. Phil. Soc., 1952. Vol. 48. P. 277-289.

50. Malinkovskiy Yu.V., Nikitenko O.A. Stationary distribution of states with bypasses and negative customers *Automatics and telemechanics*, 2000. No 8. P. 79-85. (in Russian).

51. Malinovskiy V.K. Probabilities of ruin when the safety loading tends to zero *Adv. in Appl. Probab.*, 2000. Vol. 32(3). P. 885-923.

52. Mandelbrot B. The Pareto-Levy law and the distribution of income *Internat. Econ. Rev.*, 1960. Vol. 1. P. 79-106.

53. Nagaev S.V. On asymptotic behavior of one sided large deviations *Theory Prob. Appl.*, 1981. T. 26, vol. 2. P. 369-372. (in Russian).

54. Nagaev S.V. Some limit theorems for large deviations *Theory Prob. Appl.*, 1965. T. 10, vol. 2. P. 231-254. (in Russian).

55. Nyrhinen H. Finite and infinite time ruin probabilities in a stochastic economic environment *Stochastic Process. Appl.*, 2001. Vol. 92(2). P. 265-285.

56. Osipova M.A. Development and investigation of queueing networks models by decomposition method of special kind / Autoreferat of candidate thesis. Vladivostok: *FES*, 2003. 19 p. (in Russian).

57. Pak S.M. Numerical experiment in Estes model Abstracts of 1-st Far Eastern Conference of students and post graduate students on mathematical modelling. Vladivostok: *FES*, 1997. P. 48. (in Russian).

58. Pavlov I.V. Sequential control and estimate of system reliability characteristcs using information by its elements *Engineering cybernetics*, 1986. No 6. P. 85–94. (in Russian).

59. Petrov V.V. Sums of independent variables. M.: *Science*, 1972. 416 p. (in Russian).

60. Pittel B. Closed exponential networks of queues with saturation: the Jackson-type stationary distribution and its asymptotic analysis *Math. Oper. Res.*, 1979. Vol. 4, No 4. P. 357-378.

61. Rabotnov Yu.N. Mechanics of deformed body. M.: *Science*, 1979. 741 p. (in Russian).

62. Results of science and technique, *Probability theory. Math. Statistics. Theoretical cybernetics*, 1983. T. 21. 180 p. (in Russian).

63. Rolski T., Schmidli H., Schmidt V., Teugels J. *Stochastic Processes for Insurance and Finance*. New York: Wiley, 1999.

64. Rozanov Yu.A. Lectures on probability theory M.: *Science*, 1986. 120 p. (in Russian).

65. Rozanov Yu.A. Random processes. Short course. M.: *Science*, 1971. 288 p. (in Russian).

66. Scheller-Wolf A. Further delay moment results for FIFO multiserver queues *Queuing Systems*, 2000. Vol. 34. P. 387-400.

67. "Seven instruments of quality" in Japanese economics. M.: *Edition of standards*, 1990. 88 p. (in Russian).

68. Shapiro A.N. On stochastic model of predator with adaptation Mathematical models in agrophysics and biology. L.: *Hydrometeoedition*, 1971. Vol. 30. P. 212-224. (in Russian).

69. Shapiro A.P. About cycles in recurrent sequences Control and information. Vladivostok: FESC USSR AS, 1972. Vol. 3. P. 96-118. (in Russian).

70. Sharkovskiy A.N. Differential equations and dynamics of populations *Preprint of Mathematical institute*. Kiev: Ukraine AS, 1982. 22 p. (in Russian).

71. Shiriaev A.N. Probability. M.: *Science*, 1989. 640 p. (in Russian).

72. Skvortsova N.N., Batanov G.M., Petrov A.E., Pshenichnikov A.A., Sarksyan K.A., Kharchev N.K. Non-Brownian Particle Motion in Structural Plasma Turbulence *Proceedings of the XXIII Seminar on Stability for Stochastic Models*.

Pamplona, Spain, 2003. P. 88.

73. Soloviev A.D. Analytical methods of calculations and releability estimates. Problems of mathematical reliability theory. M.: *Radio and communication*, 1983. 376 p. (in Russian).

74. Talalaeva A.B. Application of ergodicity properties and decomposition in estimate of parameters in some applied probability models *Candidate thesis. Vladivostok: FETSU*, 1997. 132 p. (in Russian).

75. Tsitsiashvili G.Sh. Quantitative Evalution of Decomposition Effects in Complex Systems *Advances in Modelling and Analysis*, 1995. Vol. 47, No 1. P. 27-30.

76. Tsitsiashvili G.Sh., Markov N.V. Asymptotic characteristics of input flows in queueing networks *Far Eastern Mathematical Journal*, 2003. T. 4, No 1. P. 36-43. (in Russian).

77. Tsitsiashvili G.Sh. Solution of mutual debts discharge *Far Eastern mathematical collection*, 1995. Vol. 1. P. 126-131. (in Russian).

78. Tsitsiashvili G.Sh. Additive corruptions of debts matrices *Preprint IAM. Vladivostok: FEB RAS*, 1996. 4 p. (in Russian).

79. Tsitsiashvili G.Sh. Coalition effects in debts discharge *Preprint IAM. Vladivostok: FEB RAS*, 1998. 6 p. (in Russian).

80. Tsitsiashvili G.Sh. Cooperative solution of barter problem *Preprint IAM. Vladivostok: FEB RAS*, 1996, 6 p. (in Russian).

81. Tsitsiashvili G.Sh. Collective insurance of large risks *Lectures of RAS*, 1999. T. 368, No 6. P. 749-750. (in Russian).

82. Tsitsiashvili G.Sh. Decomposition methods in problems of stability and effectiveness of complex systems. *Vladivostok: FEB USSR AS*, 1989. 117 p. (in Russian).

83. Tsitsiashvili G.Sh. Investigation of nonstationary commutative effects in simplest probability models *Preprint IAM. Vladivostok: FEB USSR AS*, 1991. 12 p. (in Russian).

84. Tsitsiashvili G.Sh. Commutative properties of reserve systems *Theory Prob. Appl.*, 1991. T. 36, vol. 4. P. 817. (in Russian).

85. Tsitsiashvili G.Sh., Osipova M.A. Stochastic control of parameter in discrete Markov process *Far Eastern Mathematical Journal*, 2002. T. 3, No 1. P. 58-60.

(in Russian).

86. Tsitsiashvili G.Sh. Quantitative estimate of joint effect in simplest multi-server queueing systems *Problems of stochastic models stability. M.: IAS*, 1988. P. 140-142. (in Russian).

87. Tsitsiashvili G.Sh. Decompositions of stationary random functions with finite correlation interval *Preprint IAM. Vladivostok: FEB RAS*, 1996. 11 p. (in Russian).

88. Tsitsiashvili G.Sh. Probability model of sliding regime Abstracts of scientifical conference *Dynamical systems: stability, control, optimization. Minsk: BSU*, 1993. P. 79. (in Russian).

89. Tsitsiashvili G.Sh. Communication effects in problem of threads wisp break *Preprint IAM. Vladivostok: FEB RAS*, 1993. 8 p. (in Russian).

90. Tsitsiashvili G.Sh. Estimate of sliding nonhomogenity in frames of probability model *Preprint IAM. Vladivostok: FEB RAS*, 1995. 6 p. (in Russian).

91. Tsitsiashvili G.Sh. Decomposition effects in probability model of sliding regime *Preprint IAM. Vladivostok: FEB RAS*, 1995. 5 p. (in Russian).

92. Uchaikin V.V. Multidimensional symmetric anomalous diffusion *Chemical Physics*, 2002. Vol. 284. P. 507-520.

93. Utkin V.I. Sliding regimes and their applications in systems with varying structure. M. *Science*, 1974. 272 p. (in Russian).

94. Vladimirov V.S. Equations of mathematical physics. M.: *Science*, 1967. 436 p. (in Russian).

95. Vsesviatskikh S.Yu., Kalashnikov V.V. Moments of rear events breaking in regenerative processes. *Theory Prob. Appl.* 1985. T. 30, vol. 3. P. 580–583. (in Russian).

96. Whitt W. The impact of a heavy-tailed service-time distribution upon the M/GI/s waiting-time distribution *Queueing Systems*, 2000. Vol. 36. P. 71–87.

97. Zeifman A.I. Quality properties of nonohomogenuos birth and death processes. Problems of stochastic models stability. M.: IAS USSR AS, 1988. P. 32-40. (in Russian).

98. Zolotarev V.M. One-dimension stable distributions M.: *Science*, 1983. 304 p. (in Russian).

99. Zolotarev V.M. Modern theory of independent random variables summation.

M.: *Science*, 1986. 416 p.

Index

A

algorithm, vii, viii,, 17, 20, 22, 28, 29, 34 36, 60, 101, 102, 109, 110, 150, 152, 159, 160

C

classification, 95, 170
convergence, 6, 10, 36, 47, 50, 61, 96, 105, 106, 107, 108, 125, 133, 134, 168, 169, 170

D

data processing, vii, viii, 101, 172, 174, 176
debts, 147, 164
decomposition, viii, 87, 97, 98, 99, 101, 116, 117, 171, 175, 176
diffusion, 36, 121
discrete Markov processes, viii
disease, 121
distribution function, 110, 111
division, 154

E

ecology, 15, 18, 20, 22, 23, 25, 26, 27, 28, 29, 30, 33, 34, 109, 110, 150, 151, 152, 153, 154, 155, 156, 157, 158, 159, 160, 161, 162, 163, 164
economics, vii, ix, 55, 147, 171, 173, 174

F

food, 49, 63, 164, 165
fuel, 121, 122, 125, 128, 130, 131, 132, 133, 134, 177

I

insurance, vii, viii, 10, 15, 18, 20, 22, 30, 35, 36, 40, 53, 55, 58, 63, 64, 67, 68, 74, 76, 88, 95, 98, 99, 102, 105, 106, 107, 108, 111, 112, 116, 121, 128, 132, 135, 137, 139, 144, 164, 165, 171, 172, 173, 175, 176, 177

L

layering method, 101, 110, 152, 174
linear systems, 102

M

Markov chains, 107, 164
matrix, 150, 151, 152, 153, 154, 155, 156, 157, 158, 159, 160, 161, 162, 163, 171, 173, 176
modelling, 147
models, vii, ix, 3, 5, 7, 9, 11, 13, 15, 17, 19, 21, 23, 25, 27, 29, 31, 33, 37, 39, 41, 43, 45, 47, 49, 51, 53, 55, 57, 59, 61, 63, 65, 69, 71, 73, 75, 77, 79, 81, 83, 85, 89, 91, 93, 95, 97, 99, 103, 105, 107, 109, 111, 113, 115, 117, 119, 121, 122, 124, 126, 128, 130, 132, 134, 136, 138, 140, 142, 144, 149, 151, 153, 155, 157, 159, 161, 163, 165, 167, 169, 173, 175, 177
Monte-Carlo methods, viii, 35, 36, 58, 67, 73, 88, 101, 176
multiserver queuing, viii, 1, 2, 6, 8, 9

N

networks, vii, viii, 11, 17, 20, 22, 26, 30, 55, 76, 88, 173, 174, 175, 176

O

operations research, vii, viii, 36, 164, 171, 172, 173, 175

P

parallelization, vii
probability, 1, 11, 22, 30
probability theory, viii, 2, 8
protocol, 23
psychology, viii, 147, 150, 164, 165, 175

Q

queuing systems, vii, viii, 1, 35, 58, 67, 68, 87,
    97, 101, 102, 107, 121, 133, 139, 147, 164,
    172, 174, 175, 176

R

reliability, 87
reliability systems, 56

S

science, ix, 175
search, 160
service, 77, 177
simulation, 113
stationary distributions, 1, 30
stochastic model, vii, ix, 1, 2, 3, 5, 7, 9, 10, 11,
    13, 15, 17, 19, 21, 23, 25, 27, 29, 31, 33, 37,
    39, 41, 43, 45, 47, 49, 51, 53, 55, 57, 58, 59,
    61, 63, 65, 67, 69, 71, 73, 75, 76, 77, 79, 81,
    83, 85, 89, 91, 93, 95, 97, 99, 100, 103, 105,
    107, 109, 111, 113, 115, 117, 119, 121, 122,
    124, 126, 128, 130, 132, 134, 135, 136, 138,
    139, 140, 142, 144, 149, 151, 153, 155, 157,
    159, 161, 163, 165, 167, 169, 173, 174, 175,
    177
subtraction, 110